Distributed Coordination Control System

Yongnan Jia

Distributed Coordination Control System

Application Scenarios and Countermeasures

知识产权出版社 有限责任公司
Intellectual Property Publishing House Co., Ltd.

Springer

Yongnan Jia
University of Science and Technology
Beijing
Beijing, China

ISBN 978-981-96-0364-0 ISBN 978-981-96-0365-7 (eBook)
https://doi.org/10.1007/978-981-96-0365-7

Jointly published with Intellectual Property Publishing House
The print edition is not for sale in China (Mainland). Customers from China (Mainland) please order the print book from: Intellectual Property Publishing House.

Translation from the Chinese language edition: "分布式协同控制系统应用场景及对策" by Yongnan Jia, © Intellectual Property Publishing House 2023. Published by Intellectual Property Publishing House. All Rights Reserved.

This Springer imprint is published by the registered company Springer Nature Singapore Pte Ltd.
The registered company address is: 152 Beach Road, #21-01/04 Gateway East, Singapore 189721, Singapore

Preface

In the “Tao Te Ching”, Lao Zi states, “The Tao gave birth to One, One gave birth to Two, Two gave birth to Three, and Three gave birth to all things.” This ancient wisdom reflects a fundamental truth: the evolution of all things progresses from simplicity to complexity, from individuals to colonies, and from individual intelligence to collective intelligence. Swarm intelligence, a central concept within artificial intelligence, embodies this evolutionary trajectory and represents the future of AI.

Swarm intelligence focuses on large-scale multi-agent systems (often called large systems, group systems, or complex systems), where intelligent collective behavior emerges from local interactions and simple rules. The distributed nature of these systems imparts key characteristics, including robustness, scalability, reliability, and flexibility. This book examines how distributed coordination control strategies, inspired by natural swarm behavior, can be designed to address the challenges of coordinating large-scale multi-agent systems.

This book presents an in-depth analysis of distributed coordination control, integrating insights from biological, physical, and control theory perspectives. It systematically explores the entire process, from the conceptualization and design of models to simulation-based analysis, theoretical derivation, and experimental validation. By abstracting a unified approach to addressing challenges in natural sciences, this work provides a comprehensive framework for solving complex interdisciplinary problems. To enhance the reader’s understanding of swarm intelligence and its practical applications, the book incorporates a variety of scenarios that illustrate the diverse strategies employed in swarm behavior, thereby bringing theoretical concepts with real-world implementations.

The book is organized around typical application scenarios of distributed coordination control strategies, drawing inspiration from swarm intelligence. It provides a thorough examination of the entire process, from mathematical modeling and theoretical derivation to simulation analysis and real-world engineering applications. The book is divided into eight chapters. Chapter 1 introduces the background and significance of the research, offering a comprehensive explanation of the motivation underlying the book’s themes. Chapter 2 traces the origins and evolution of swarm intelligence, emphasizing its manifestations in both natural systems and human society. It also analyzes the fundamental characteristics of swarm behavior, clarifying the core principles of swarm intelligence and distributed coordination. Chapter 3 delves into the concepts and major research directions of distributed coordination control systems, outlining the essential theoretical tools and key technologies central to this filed of study. Chapters 4–7 focus on various types of distributed coordination control systems. Each chapter addresses different system configurations, ranging from systems with homogeneous agents to those with constant-speed or variable-speed leaders, and presents detailed solutions tailored to each system. Chapter 8 explores optimization strategies for distributed coordination control systems, drawing inspiration from the collective behavior of social animals. It investigates critical aspects such as convergence stability and speed, offering practical insights for enhancing system performance.

It is crucial to highlight that scientific inspiration is often derived from life itself, and the ultimate aim of research is to benefit life. Guided by this principle, the author has structured this book with the hope that it will inspire readers to uncover the beauty inherent in life, to pursue the guiding light of existence, and to appreciate the profound wisdom embedded in our survival. This work seeks to provoke critical reflection, urging readers to explore the logic of swarm intelligence and to transcend traditional disciplinary boundaries, in pursuit of the unified mechanisms that govern interdisciplinary collaboration.

In his seminal work *Engineering Cybernetics*, Qian Xuesen pioneered the application of cybernetics to engineering, thereby opening new avenues for research in control theory. Inspired by Qian's contributions, this book offers two primary contributions: First, it presents a comprehensive research framework for distributed coordination control based on swarm strategies, covering the entire process from theoretical derivation and simulation analysis to experimental validation. Second, it presents the first attempt to integrate rigorous stability analysis of complex systems with statistical methods, leading to novel and thought-provoking insights. While swarm intelligence research remains in its early stages, this book aims to make a collective exploration of the future of swarm intelligence.

Finally, it is important to acknowledge that this book is based on the research outcomes of the National Natural Science Foundation of China's Youth Project, "Three-Dimensional Cooperative Control of Manned/Unmanned Aerial Vehicle Swarms" (No. 61603362), as well as collaborative research conducted with Prof. Tamas Vicsek, a member of the Hungarian Academy of Sciences.

Beijing, China Yongnan Jia

Contents

Chapter 1
Collective Phenomenon

Swarm phenomena are ubiquitous in both nature and human society.

In nature, populations often exhibit spatial clustering, forming groups of varying sizes to utilize space efficiently. For example, many marine fish display clear swarm behavior during spawning, foraging, and migration, with the shape and size of these swarms varying by species. Similarly, phenomena such as public opinion events in dense crowds or synchronized clapping in large gatherings reflect the principles of swarm behavior. Over years of evolution, humans, as intelligent beings at the top of the food chain, have also realized the importance of cooperation and collective strength to gain a competitive edge.

From an ecological perspective, adopting swarm strategies provides significant survival advantages for organisms, especially those that cannot survive independently. Firstly, swarming facilitates mating and reproduction among individuals. Secondly, swarms offer mutual protection; when faced with a predator, fish can quickly form a defensive group, with lone individuals are more likely to fall prey. Thirdly, group foraging is advantageous; when part of a fish school encounters a favorable feeding area, it remains there while the rest of the group circles to ensure everyone benefits. Ants employ similar strategies. Fourthly, group movement creates favorable dynamic conditions, reducing resistance compared to solitary movement and maximizing swimming efficiency. Fifthly, swarming can alter the chemical properties of the environment; studies have shown that fish in swarms are better at resisting toxins than when living individually. However, swarm strategies can also present challenges, such as increasing the likelihood of being targeted by predators due to the large group size, risks of competition for food, and the spread of disease within the group.

Have you ever marveled at the breathtaking sight of a murmuration of millions of starlings, painting the evening sky with ever-changing, fairy-tale-like patterns? Have you ever imagined yourself as a free fish swimming in the ocean, moved by the sight of a sardine school forming a massive ball to fend off a shark attack? Do

Y. Jia, *Distributed Coordination Control System*,
https://doi.org/10.1007/978-981-96-0365-7_1

you wonder what secrets lie behind such spectacular large-scale collective behavior? This chapter will unveil the mysteries behind swarm phenomena, layer by layer.

1.1 Swarm Phenomena in Nature

Swarm phenomena are prevalent across the natural world. In many species, individuals form groups of varying sizes as a strategy for spatial organization and survival. For instance, numerous marine fish exhibit pronounced swarming behavior during critical life events such as spawning, foraging, and migration. The shapes and sizes of these swarms can vary significantly depending on the species. Beyond the marine environment, swarm behavior is also observed in terrestrial animals, such as the synchronized movements of bird flocks or insect colonies.

Swarm phenomena typically emerge in biological groups composed of self-driven individuals. Although the capabilities of a single organism—such as movement, perception, and vision—are inherently limited, the collective behavior of a group, whether small or large, can exhibit complex and coordinated dynamics. This is primarily due to the information exchange between neighboring individuals within the population, enabling the group to function as a cohesive unit despite the limitations of each member.

1.1.1 Typical Swarming Phenomena

As shown in Fig. 1.1, collective behavior is observed not only in living organisms, such as bacterial aggregation, seasonal bird migrations, fish schooling and migration, herd movements, bee social behavior, and ant foraging, but also in non-living entities, such as molecules and galaxies. In a school of fish, each fish uses its vision and lateral line system to sense changes in water flow, aiming to synchronize its speed and direction with other fish in the group. This efficient collective behavior plays a crucial role in the fish's search for food, migration, and evasion of predators, allowing the school to move toward a common goal at high speed while effectively avoiding predators or obstacles [1]. Additionally, hazardous collective behaviors, such as locust plagues, also exist in nature [2].

1.1.2 Advantages of Collective Strategies

In general, collective phenomena refer to the self-organized decision-making of individuals in a group based on local information exchanges with their neighbors, leading

(a) archaea, virus, bacteria, fungus. (b) cattle nose gansu migration. (c) fish schooling. (d) cattle migration. (e) swarm of bees. (f) ant colony. (g) flocks of birds. (h) galaxy. (i) cells.

Fig. 1.1 Swarm Phenomena in Nature

to emergent coordinated behaviors or the formation of a cooperative, freely organized group capable of completing complex tasks that individual members cannot accomplish alone.

From a biological perspective, collective strategies offer numerous advantages to animal groups. For example, during collective foraging, cattle egrets can disturb ground-dwelling insects, making it easier to capture prey [3]. Lions can hunt large animals such as zebras and buffaloes; cranes can extend foraging time as the group size increases; sparrows and white-tailed deer use group dynamics to better detect predators, with individuals signaling each other through sounds, behaviors, and chemical cues to evade threats and reduce predation risk; white bass and violet-backed starlings confuse predators and decrease predation probability through large-scale group movements; herons, ibises, and some crow species benefit from collective defense against predators; meerkat groups provide better care for their young as group size increases, improving offspring survival rates; emperor penguins huddle together to stay warm and incubate eggs during the harsh winter, enduring the long season. In summary, collective behavior provides at least four key advantages: first, reducing the risk of predation; second, decreasing the difficulty of foraging; third, enhancing mating efficiency; and fourth, improving overall defense mechanisms [4–6]. However, collective strategies also have drawbacks, such as increased competition and disease transmission risk in the presence of limited food resources. Nevertheless, the collaborative mechanisms underlying collective behavior offer valuable insights for scientists and engineers. Thus, it is essential to fully understand collective phenomena and explore the collaborative control mechanisms behind them to better guide and benefit human activities.

1.1.3 Confusions About Collective Strategies

As early as two thousand years ago, the Roman naturalist Pliny observed collective behavior in starling flocks. In the 1960s, Frank Heppner, an emeritus professor and ornithologist at the University of Rhode Island, began studying starling behavior. Despite over fifty years of research, the highly coordinated movements and flight patterns of starlings still puzzled him.

The fascination with collective phenomena also extends to Giorgio Parisi, who won the Nobel Prize in Physics in 2021. One day, he observed thousands of starlings flying in unison over a Roman train station. Why do starling flocks, without a centralized command, manage to change shape like a single entity? Driven by this question, he collaborated with physicist Andrea Cavagna from the University of Rome and others to set up three synchronized high-speed cameras on the roof of Palazzo Massimo to capture long-term footage of starling flocks during their winter migration. Using the footage, they reconstructed the three-dimensional coordinates of starling movements and simulated the positions and speeds of individual starlings in a computer model, creating a complex 3D model of the flock. They then applied statistical physics methods to analyze how millions of starlings form a cohesive whole.

The experimental data revealed that although starling flocks exhibit complex and variable shapes, the mechanism maintaining the flock is simple: individuals adjust their behavior based on the actions of their seven nearest neighbors, leading to complex formations at the group level. When one bird changes speed and direction, nearby individuals quickly follow and coordinate their changes. Information spreads rapidly throughout the large flock. Although each bird is influenced only by its neighboring birds, each bird's movement ultimately affects and is affected by the entire flock. This results in a constant speed of information transmission within the flock. As a result, the speed of collective decision-making is so rapid that a turning signal from a peripheral starling can spread to a flock of 400 starlings in half a second (at a speed of about 145 km/h).

The various shapes and patterns formed by starlings in the sky are given evocative names such as "bubbles," "alert lines," and "flash expansions." Charlotte Hemelrijk, a professor of evolutionary life sciences at the University of Groningen in the Netherlands, notes that one of the most fascinating sights is when a large flock of starlings changes shape rapidly like a flowing cloud, with black ribbons forming between the dispersed and dense areas of the flock.

The reason for the flocking of European starlings into such large groups in the sky is puzzling. The most common explanation is "safety in numbers," meaning that collective behavior is a protective response to predators. Initially, it might be thought that starlings prefer to shorten their flight time, but in reality, such spectacular displays often last 30–45 minutes, during which starlings expend energy at an astonishing rate. Simultaneously, such behavior actually attracts the attention of predators, increasing their risk of exposure.

Another possible explanation is the "thermoregulation" theory. This theory suggests that the vocalizations serve to advertise the roosting site, attracting more starlings to the flock for warmth. To unravel this mystery, researchers from the University of Gloucestershire and the Royal Society of Biology compiled data from over 3,000 starling flocks collected by volunteers from 23 countries in 2014 and 2015. A paper published in 2017 showed that there is no correlation between temperature and flock size, undermining the thermoregulation hypothesis. The study also found that data on the presence of hawks, falcons, or other raptors in starling flocks accounted for less than a third of the cases, providing some support for the predation theory but still not fully explaining why starlings engage in such lengthy aerial displays.

1.1.4 Insights from Collective Phenomena

The allure of collective phenomena has attracted significant scholarly attention, leading to some widely accepted research findings. Collective strategies, also known as "scale-free behavioral correlation," exhibit interesting "critical" characteristics. "Critical" refers to a state that lies precisely between "ordered" and "disordered," similar to the critical point in phase transitions. At this point, a group can maintain its stability while ensuring effective information transfer within the group. Based on the aforementioned collective phenomena, we can identify the following common advantages of collective strategies:

First, collective strategies exhibit rapid response times, with information spreading quickly within the group and maintaining this speed even as the group size increases, as seen in species like the wandering bat, European starling, and locusts.

Second, collective strategies often emerge as survival tactics when individual organisms cannot cope alone, such as in predator defense, long-distance migration, and reproduction.

Third, collective behavior typically features distributed characteristics, with each individual exhibiting self-driven properties, leading to significant flexibility. Examples include the alternating shifts of emperor penguins huddling for warmth and the rotation of lead geese in migration.

Fourth, collective strategies are particularly suited for large-scale and super-large-scale group coordination, with individuals avoiding collisions, reflecting the exquisite craftsmanship of nature's evolution.

1.2 Collective Phenomena in Human Society

Collective phenomena are not only widespread in the natural world but also common in human society. For example, these phenomena can be observed in synchronized applause by crowds, gatherings and dispersals of people, and traffic flow congestion and management. Collective behavior in human society, also known as collective or mass behavior, refers to actions that occur spontaneously among large groups of people without formal guidance, specific goals, or adherence to established social norms. A frequently mentioned characteristic of collective behavior is the abandonment of individual beliefs and attitudes in favor of a new collective behavioral pattern. Despite the diverse backgrounds of group members, their behavior within the collective often confirms to a new set of shared norms, resulting in a unified collective spirit.

1.2.1 From Collective Behavior to Distributed Coordination: Crowd Evacuation

In reality, humans live in clusters in both urban and rural areas. Additionally, many activities in human society exhibit collective behavior. For instance, crowded stampedes frequently occur during large-scale events both domestically and internationally. Since 2000, over 100 stampede incidents at major events worldwide have been reported, resulting in nearly 6,000 deaths and about 10,000 injuries, causing severe social impacts. In 2011, a magnitude 9 earthquake in Japan triggered a tsunami approximately 10 meters high, causing nearly 16,000 deaths and over 2,500 missing persons. In 2019, China experienced over 230,000 fires, resulting in more than 2,000 casualties and direct property losses amounting to 3.6 billion yuan. During natural and man-made disasters such as earthquakes, fires, and stampedes, effective crowd evacuation becomes a critical issue. By leveraging the mechanisms of collective behavior to influence, adjust, or even control people's actions and social activities, many tragedies can potentially be prevented.

During the 2008 Wenchuan earthquake, 2,300 teachers and students at Sangzao Middle School, located in the most severely affected area, completed an evacuation plan within one minute and orderly moved to the playground. The Sangzao Middle School case applied the concept of distributed coordination from collective behavior, having rehearsed the evacuation plan multiple times before the earthquake. This approach effectively controlled the direction and speed of the crowd movement in densely populated areas, maintaining behavioral consistency and significantly increasing evacuation speed, thereby greatly reducing casualties and property damage.

1.2.2 *From Collective Behavior to Distributed Coordination: Traffic Congestion*

With the continuous development of human society, expanding urban areas, and increasing population density, traffic congestion has become a severe issue, significantly raising travel costs. By the end of 2020, Beijing's vehicle ownership had reached 6.57 million, with private small and micro cars accounting for 4.73 million. On weekdays, traffic congestion exceeded two hours. According to statistics from the Chinese Academy of Social Sciences, the social cost of traffic congestion increases by approximately 40 million yuan daily, with total losses exceeding 14 billion yuan annually. Traffic congestion not only prolongs travel time but also increases fuel consumption, impacting daily life and contributing to issues such as exhaust and noise pollution.

Applying the concept of distributed coordination from collective behavior to address these social problems could significantly reduce personnel and property losses, benefiting society. Many countries have begun to apply collective principles to urban traffic control. For example, the U.S. Department of Transportation has implemented the "National Intelligent Transportation System Program," which has reduced traffic congestion by 20%. This program includes measures such as controlling traffic flow, regulating vehicle behavior, sharing traffic information, and real-time transmission and application of traffic data. Intelligent transportation systems are now implemented in over 80% of traffic facilities. In 2016, the European Union adopted the "European Cooperative Intelligent Transport Systems Strategy," aiming to widely promote cooperative intelligent transportation systems. This strategy seeks to achieve "smart communication" between vehicles and road infrastructure, as well as among vehicles, with the goal of reducing road traffic fatalities by half from 2010 to 2020.

1.3 Summary

This chapter introduces collective phenomena in both the natural world and human society. Studying these phenomena offers profound scientific value in understanding nature; exploring collective concepts provides important insights into human social patterns and development trends; and uncovering the underlying mechanisms of collective behavior holds broad application prospects for solving complex system engineering challenges.

References

1. Vabo, R., & Nottestad, L. (1997). An individual-based model of fish school reactions: Predicting antipredator behaviour as observed in nature. *Fisheries Oceanography, 6*, 155–171.
2. Couzin, I. D., & Krause, J. (2003). Self-organization and collective behavior in vertebrates. *Advances in the Study of Behavior, 32*, 1–67.

3. Scott, D. (1984). The feeding success of cattle egrets in flocks. *Animal Behavior 32*(4), 1089–1100.
4. 余玉群，刘楚光，郭松涛，等.天山盘羊集群行为的研究[J].兽类学报，2000, 20(2): 101-107. (in Chinese)
5. Foster, W. A., & Treherne, J. E. (1981). Evidence for the dilution effect in the selfish herd from fish predation on a marine insect. *Nature, 293*, 466–467.
6. Olson, R. S., Hintze, A., Dyer, F. C., & Knoester, D. B., & Adami C. (2013). Predator confusion is sufficient to evolve swarming behavior. *Journal of the Royal Society Interface, 10*, 20130305.

Chapter 2
Collective Systems, Collective Intelligence, and Collective Strategies

John Downer, the producer of the *Life in the Wild* series, states: "Nature creates intelligent behavior through two evolutionary paths. One is by individual organisms developing large and complex brains, like humans. The other is by forming groups of millions of small-brained organisms that communicate with each other, such as swarms."

In nature, numerous examples of collective behavior exist. Observations and studies reveal a common feature: these group behaviors exhibit intelligent actions through communication and cooperation among individuals. Each individual adjusts its actions by observing the behavior of others in the group. Compared to individual behavior, group behavior offers several advantages, such as faster food acquisition and better predator avoidance.

As research into group behavior in both natural and human societies has advanced, it has become clear that the cooperative mechanisms observed in these behaviors can be applied to engineering practices. This has led to the study of swarm systems, swarm intelligence, and swarm strategies. By analyzing, modeling, and controlling swarm systems, researchers can explore the intrinsic principles and mechanisms of collective phenomena. This deeper understanding provides valuable theoretical guidance and technical support for managing complex systems with large scales, diverse objectives, and intricate sub-tasks.

2.1 Swarm Systems

Swarm systems generally consist of numerous small individuals that cooperate and self-organize, following specific rules to accomplish complex tasks that would be difficult or even impossible for a single individual to achieve alone. In nature, swarm behavior is often a dynamic process, where trade-offs are made to maximize benefits while adapting to the ever-changing environment.

Y. Jia, *Distributed Coordination Control System*,
https://doi.org/10.1007/978-981-96-0365-7_2

Studying the collective behavior of these organisms not only enhances our understanding of biological behavior but also enables us to apply their efficient, collaborative actions to various human endeavors. For example, leveraging animals' high synchronization, precise information transfer, and collision avoidance can be applied to control drone swarms and satellite constellations for optimal scheduling. Additionally, understanding and predicting locust swarm behavior-where high density and mutual interaction lead to organized groups-can help in controlling locust plagues.

Inspired by collective phenomena in nature (such as animals, plants, and microorganisms) and human society, swarm systems aim to replicate and optimize multi-agent behaviors. The core advantage of these systems is their ability to achieve high performance, reliability, and flexibility at a relatively lower cost. A deeper exploration of swarm systems offers significant social and economic benefits, supporting strategic opportunities in artificial intelligence development, enhancing early advantages in AI, and accelerating the creation of a technologically advanced nation.

Swarm systems are characterized by a large number of intelligent agents integrated within an open architecture. Centered around communication networks, these systems rely on emergent group intelligence, collaborative interaction between platforms, and the task execution capabilities of individual platforms. They offer advantages such as resilience, low cost, and distributed functionality.

2.2 Swarm Intelligence

Swarm intelligence originates from the widespread collective behavior of organisms in nature and is closely related to fields such as economics, sociology, management, control engineering, and computer science. Although there is no unified definition, swarm intelligence generally refers to the intelligent behavior exhibited by a group of many low- or non-intelligent individuals through simple cooperation among themselves.

Swarm intelligence has the following characteristics.

(1) **Distributed Control**: There is no central control, making it more adaptable to networked environments and enhancing robustness. The failure of one or a few individuals does not significantly affect the swarm's ability to solve the overall problem.
(2) **Indirect Communication**: Each individual can alter the environment, serving as a form of indirect communication known as **stigmergy**. This type of communication results in minimal communication overhead as the number of individuals grows, providing good scalability.
(3) **Simple Individual Rules**: The abilities or behavioral rules followed by each individual are simple, making the implementation of swarm intelligence convenient and straightforward.
(4) **Emergent Intelligence**: Complex behaviors emerge from the interactions of simple individuals, showcasing self-organization. Swarm intelligence operates

through emergence, guided by appropriate evolutionary mechanisms-something that individual intelligence alone cannot achieve.

The rapid development of artificial intelligence presents new opportunities and challenges for research in swarm intelligence. The 2017 "New Generation Artificial Intelligence Development Plan" issued by the State Council highlights swarm intelligence theory as one of the eight foundational theories of the new generation of artificial intelligence(AI). It emphasizes research in swarm intelligence structure theory, organizational methods, incentive mechanisms, emergence mechanisms, learning theories, and general computing paradigms and models for swarm intelligence.

The plan also identifies key swarm intelligence technologies as one of the eight critical common technologies for the new generation of AI, and swarm intelligence service platforms as one of the five foundational support platforms for this generation. The focus includes proactive perception, knowledge acquisition and generation, collaboration and sharing, evaluation and evolution, human–machine integration and enhancement, and self-maintenance and secure interaction. Additionally, it aims to establish service systems for swarm intelligence space, research collaborative decision-making and control technologies for mobile swarm intelligence, and develop various support platforms for swarm intelligence.

The "Technology Innovation 2030—'New Generation Artificial Intelligence' Major Project 2018 Annual Project Application Guide" issued by the Ministry of Science and Technology also specifies that swarm intelligence is one of the five continuous research directions in AI. As an important area of research in complex systems and AI, swarm systems have extensive application prospects in aerospace, national defense, transportation, agriculture, and other fields.

2.3 Swarm Strategies and Distributed Coordination

2.3.1 Core Ideas of Swarm Strategies

Swarm phenomena have attracted widespread attention across various fields, including biology, physics, control engineering, computer science, and sociology. Scholars in these areas use tools such as modeling, simulation, and derivation to explore how groups can maintain coordinated behavior despite lacking global information and facing external disturbances.

In swarm systems, self-driven individuals rely on local information exchange and simple rules, eventually exhibiting coordinated and synchronized group motion. This type of swarm intelligence, which emerges from local behaviors without central control, significantly surpasses the capabilities of individual agents. It enhances survival, predation, and other abilities of biological groups, maximizing benefits for the entire population. The core idea of swarm intelligence can be summarized as leveraging scale advantages to achieve high performance, reliability, flexibility, and scalability at relatively low costs.

The concept of swarm intelligence holds significant engineering value:

(1) **Low Cost**: Swarm strategies require minimal intelligence from the agents, relying only on simple data processing, environmental perception, and information exchange among agents.
(2) **Scalability**: The scale-invariance characteristic allows swarm strategies to be easily extended from small to large groups without exponential increases in computational and communication demands.
(3) **High Performance**: Swarm behavior emerges from local interactions and distributed architecture, resulting in collective behaviors that far exceed individual capabilities, illustrating the emergence of swarm intelligence.
(4) **Robustness**: Due to self-organizing properties, swarm systems can continue to perform complex tasks even if some agents fail, exemplifying the core strength of swarm intelligence.

2.3.2 Core Concepts of Distributed Architecture

Distributed systems, as an organizational form, contrast with centralized systems and are prevalent in various fields. For example, distributed computing in computer science breaks down complex problems into smaller parts, which are processed across multiple computers, with results integrated to form the final outcome. Similarly, distributed network storage uses a scalable architecture where storage servers share the load, and location servers help locate stored data. These systems overcome the bottlenecks of traditional centralized computing, enhancing reliability, availability, and scalability.

A distributed control system (also called a decentralized control system) manages production processes with centralized management and decentralized control. This system emerged in response to the increasing complexity and automation levels in modern industrial production. By integrating computer technology, network technology, communication, and automatic control, distributed control systems optimize the distribution of risk and centralized management. They employ decentralized control principles and hierarchical structures, and are widely used in industries like petroleum, chemicals, electricity, metallurgy, and smart buildings.

The core idea of a distributed organizational structure involves breaking down a complex task or organization into sub-tasks or sub-units, with coordination and collaboration between them to achieve the overall goal. This approach significantly enhances system reliability, scalability, performance, and flexibility.

In this book, we focus on distributed coordination control systems, specifically studying multi-agent systems with distributed organizational structures designed to achieve collective coordination.

2.3.3 *Unity of Swarm Strategies and Distributed Coordination*

The octopus is one of the most "enigmatic" animals on Earth and among the most intelligent invertebrates. It possesses a vast number of neurons, with 60% distributed across its eight arms and only 40% in its brain. As a result, its arms have independent thinking capabilities and exhibit agile responses. During hunting, the octopus is exceptionally quick and dexterous, with excellent coordination between its arms, which never get tangled. This arrangement resembles a distributed computing model of "multiple small brains + one big brain." Utilizing this distributed empowerment model, Professor Huang Yuchang from China Europe International Business School developed a new chain operation mechanism for BBK Electronics, assisting the company in its digital transformation and rapid scaling. If each franchise store is viewed as a self-driven entity, then the digital transformation of the enterprise can be seen as designing operational strategies for each store to maximize the overall profit margin of the company. In-depth research on swarm strategies can provide better optimization solutions and deeper theoretical support for this digital transformation.

Professor Huang believes that digitalization involves constructing an open, multi-dimensional collaborative network. By implementing finer and more diverse division of labor and adopting a looser organizational structure, strong innovation capabilities are generated. The organizational form of enterprises has shifted from a centralized, tree-like structure to a distributed, networked ecological architecture. The greatest benefit of this organizational form is not only its strong innovative power but also its ability to cope with external uncertainties and risks. This reflects the advantages of swarm strategies and distributed architectures.

Clearly, the idea of distributed coordination holds significant value not only in engineering applications but also in management, sociology, and other fields. Especially in the context of the current digital transformation, this has profound implications.

It is important to emphasize that the term "swarm" here specifically refers to the phenomenon where self-driven individuals in nature or human society exhibit complex collective behaviors through local information exchange and simple rules, rather than merely being gathered together.

Researching swarm issues essentially involves studying the emergence of swarm intelligence. In this book, swarm control problems and distributed coordination control problems can be considered the same type of issue. Both address complex systems composed of multiple agents (or organizations, institutions, or tasks), with the former focusing more on achieving specific outcomes and the latter on the organizational forms that enable problem-solving. Alternatively, swarm control problems can be viewed as a special category of distributed coordination control problems, concentrating on designing control protocols based on biological swarm concepts to address distributed coordination tasks.

2.4 Summary

This chapter introduced the definitions and characteristics of swarm systems and swarm functions, outlined the core principles of swarm strategies, and explored the relationship between swarm strategies and distributed coordination. Practical examples were used to illustrate the coordination and integration of these core ideas. The emphasis on swarm systems in this chapter highlights their significance in solving distributed coordination control problems and their broad applicability in fields such as engineering, business management, commercial strategy, and public opinion control.

Chapter 3
Distributed Coordination Control System

Scientists and researchers have long advanced scientific knowledge by observing phenomena in the natural world, resulting in many remarkable technological achievements.

In the field of biology, swarm behavior has been a focal point of attention. Different biological entities spontaneously gather in space to act as a cohesive unit or form coordinated synchronous movements. Examples include the migration of birds in flocks, the collective swimming of fish in the ocean, cooperative hunting by mammals, and collective foraging by insects and microorganisms [1–3]. These macro-level, orderly collective behaviors exhibit swarm intelligence, a trait not found in individual organisms [4]. Swarm intelligence enables individuals to achieve benefits in activities such as foraging, mating, and avoiding predators that would not be possible through solitary actions.

Vivid examples also exist in human society, such as crowd panic and evacuation, cooperation in multiplayer games, and the emergence of fashion trends. Through the transmission and interaction of information between local individuals, these activities demonstrate efficient collaboration and advanced intelligence at the collective level.

As described above, swarm behavior is a type of collective action among social organisms, where individuals can only perceive the information of their nearby neighbors. In the absence of centralized coordination and control mechanisms, how do these individuals achieve coordinated group behavior in tasks such as foraging and predator avoidance through local coordination rules and control strategies? This chapter focuses on how to utilize swarm intelligence to address distributed cooperative control problems in multi-agent systems.

Y. Jia, *Distributed Coordination Control System*,
https://doi.org/10.1007/978-981-96-0365-7_3

3.1 Multi-agent System

The phenomenon of swarm behavior in nature has inspired the emergence and development of multi-agent systems. With the rapid advancement of technologies such as computer networks, communication, and sensors in recent years, research on intelligent mobile robots and unmanned aerial vehicles has become a frontier in scientific research. Multi-agent systems have garnered widespread attention from experts and scholars worldwide. By designing distributed cooperative control algorithms for multi-agent systems, it is possible to accomplish complex tasks that a single agent cannot achieve, with important applications in various fields such as ocean resource exploration, large-scale disaster search and rescue, swarm warfare, social services, public safety, and defense equipment upgrades.

The concept of multi-agent systems has developed along different research paths in various fields, and it is the result of interdisciplinary integration, including biology, computer science, systems control science, and physics. Currently, there is no strict definition of an agent or a multi-agent system. Generally speaking, an agent (also referred to as an individual) can represent a physical or abstract entity. An agent should possess certain autonomous capabilities, such as the ability to perceive external information, process the acquired information, and self-regulate. Multi-robot systems, sensor networks, and clusters of artificial satellites all fall under the category of multi-agent systems.

In the field of computer science, the concept of an agent was introduced in 1986 by Marvin Minsky, a founding figure of artificial intelligence and a renowned computer scientist at MIT, in his book *The Society of Mind* [5]. Each agent possesses the ability to interact with other agents and, ultimately, with humans (end operators). The interaction between different agents involves both cooperation and competition: cooperation refers to different agents coordinating with each other to complete tasks that a single agent cannot accomplish, while competition involves different agents vying for limited computing and communication resources during the coordination process. A group of rational individuals and their interactions thus form a multi-agent system. Due to its high flexibility and adaptability, multi-agent systems have gradually become a research hotspot in the field of distributed artificial intelligence [6, 7].

In control science and engineering, the field has evolved from controlling individual systems (e.g., proportional-integral-derivative (PID) control, adaptive control, etc.) to managing multiple interconnected systems. In recent years, with the rapid development of technologies such as artificial intelligence, computers, communication, sensors, and microelectronics, the basic structure and operational mode of control systems have undergone fundamental changes. The individual components of the system are no longer just controlled objects, sensors, or controllers with single functions; they have evolved into agents with sensing, computing, execution, and communication capabilities. The exchange of information among agents facilitates the coordination and cooperation process, replacing the entire control and decision-making process. Communication and collaboration play an increasingly important role. Such systems do not pursue large, complex structures but instead rely on inter-agent communication and coordination to collectively complete the collaborative control tasks of complex systems, thereby enhancing the system's robustness, reliability, and self-organization capabilities.

From the perspective of systems and control theory, the study of multi-agent systems has several novel characteristics: First, multi-agent systems structurally feature "individual dynamics + communication topology," meaning that each agent within the system adjusts its dynamic behavior through information exchange with other agents. Therefore, the overall dynamics of the system are determined by the dynamics of each agent and the communication topology among all agents. Second, the dynamics of multi-agent systems arise from simple individual behavioral rules and local information, meaning that collective behavior is an emergent, self-organized motion resulting from the cooperation among all agents. Third, in terms of control objectives, the basic task of coordinated control in multi-agent systems is to achieve a desired system configuration and overall motion pattern, such as the consensus problem, formation control problem, rendezvous problem, swarming problem, and attitude alignment problem.

Reviewing research on multi-agent systems across different disciplines, we find that despite differences in research approaches, modeling methods, motivations, and objectives, multi-agent systems across various fields share common characteristics, including autonomy, distributed nature, and evolution. Autonomy indicates an agent's ability to self-manage and self-regulate, which reflects the agent's capacity to automatically adjust its state and behavior according to its constantly changing external environment, rather than passively responding to external stimuli. The distributed nature of agents indicates their ability to collaborate with other agents or humans to solve problems. Evolution refers to an agent's learning ability, enabling it to accumulate experience and knowledge continuously to optimize its behavior and adapt to changing external environments. The two fundamental elements of multi-agent systems are the group's overall objectives and behaviors, as well as the local interaction mechanisms and individual behavior patterns.

Summarizing the current research on multi-agent systems, scholars mainly focus on system analysis, system modeling, and system control. In the analysis of multi-agent systems, researchers primarily study the impact of communication topology among agents on the system's overall state and task performance [8]. These impacts include: the effects of communication among agents, such as noise, packet loss, asynchrony, and time delay in communication channels affecting system performance in practical applications; the influence of topological structure, where researchers examine the conditions that the communication topology must meet to achieve coordinated control; and the effects of external forces, analyzing how the final state of a multi-agent system is disturbed by external signals, where multi-agent systems typically correspond to models involving external forces. In system modeling, scholars often model the communication topology among agents and the dynamic model of individual agents based on the specific characteristics of the multi-agent system under study to conduct in-depth analysis and research. In modeling, the communication topology among agents can be classified as unidirectional or bidirectional, switching or fixed, with or without time delays; based on the dynamics of agents included in the multi-agent system, the system can be classified as continuous-time or discrete-time, nonlinear or linear.

Currently, the modeling of multi-agent systems can be broadly divided into several categories [9]: repulsion-attraction interaction models [7], particle swarm models [10], Boid models [11], networked systems and graph-theoretic models with external forces [12]. Regarding the cooperative control of multi-agent systems, researchers have published numerous studies, which can be summarized into the following categories:

1. **Flocking Problem**: The flocking problem originated from the well-known Boid model proposed by Craig Reynolds. In the Boid model, each individual follows three basic rules during flight: collision avoidance, velocity matching, and flock centering. These three rules are often simply referred to as separation, alignment, and cohesion, and they succinctly summarize the basic characteristics of most group movements in nature. The goal is to ensure that all agents converge on the same velocity, maintain a stable distance between them, and avoid collisions. Based on these three basic rules, researchers have designed flocking control algorithms to meet different requirements and conditions [13].
2. **Swarming Problem**: Swarming control requires that all individuals in the group converge and remain within a bounded range centered on the group's weighted center. In nature, swarming behavior is ubiquitous; for example, when South African sardines migrate in groups and encounter sharks, they spontaneously gather to form a large "bait ball" to protect themselves from attacks. Studying the swarming behavior of multi-agent systems is a way to understand the complexity of biological systems, which may explain the emergence of swarm intelligence. Furthermore, we can apply the wisdom of biological swarming behavior to the coordinated control of multi-agent systems by using distributed strategies based on neighbor information. The swarming problem requires that each individual in the multi-agent system converges within a bounded range centered on the weighted center of the swarm and remains stable [14].
3. **Containment Control Problem**: The containment control problem involves controlling multiple agents according to certain control laws, where a subset of agents forms a convex hull formation, and the remaining agents enter and stay within the convex hull. For example, in nature, female silkworms intermittently emit hormones to attract male silkworms, and the male silkworms eventually enter the convex hull formed by the female silkworms. When a group of agents migrates from one location to another destination, some agents have strong detection, reception, processing, and feedback capabilities, while others are weaker in these aspects. To resist external threats, well-equipped agents form a convex hull around the perimeter, while the less equipped agents remain in the safe area formed by the well-equipped agents [15].
4. **Consensus Problem**: The consensus problem studies how to design control laws so that certain state variables of agents in a multi-agent system converge to the same value and achieve consensus. It requires that agents in the system respond

to changes in their environment with corresponding behavioral adjustments. To achieve consensus, each agent must follow certain rules, meaning a "consensus algorithm" (or consensus protocol) must be designed for each agent. If the designed consensus algorithm is effective, a group of agents will respond to environmental changes according to the consensus algorithm and reach an agreement on "what has changed." For example, a school of fish swimming in the same direction shares information about the swimming direction. If a fish in the front detects a potential threat or a new food source and changes its direction, the fish behind will adjust their direction based on the direction of their neighbors, and the entire school will reach a new consensus on the swimming direction [16].

5. **Formation Control Problem**: The formation control problem refers to the scenario where all agents interact with their neighbors to form a specific geometric pattern, either without a reference signal or virtual leader, or with the guidance of a reference signal or virtual leader. In multi-agent systems, formation control without a reference signal or virtual leader is referred to as "Formation Producing," while formation control with a reference signal or virtual leader is known as "Formation Tracking." For example, studies in [17–19] focus on formation producing, while the research in [20] addresses formation tracking.

As a significant branch within the field of artificial intelligence, multi-agent systems have gradually become a hot topic in various disciplines. This rise in interest is closely linked to the rapid advancements in artificial intelligence and control theory, alongside the relentless pursuit and exploration of modern science by researchers.

It is important to note that, in general, the swarm control problem refers to the clustering of individual agents, while the flocking problem not only requires agents to maintain a dense and stable formation but also to keep the same speed and direction. In this book, the goals of swarm control and flocking control are considered to be aligned.

3.2 Distributed Coordination of Multi-agent System

The analysis and research of multi-agent systems primarily focus on examining the dynamical characteristics of agents within the system and how the communication topology formed between agents influences the system's final motion state. Multi-agent systems typically employ a distributed architecture, relying on information exchange networks with the goal of forming a coordinated overall behavior. Therefore, swarm systems or narrowly defined multi-agent systems can also be referred to as distributed cooperative control systems. While multi-agent systems are defined from the perspective of the controlled objects in a complex system, distributed cooperative control systems are defined from the control methods' perspective, and swarm systems are defined from the task's perspective. In this book, we will focus on multi-agent systems, using the dynamical characteristics of agents within the system as constraints and drawing inspiration from swarm phenomena (strategies) to elucidate the distributed cooperative control problems of multi-agent systems.

Distributed cooperative control systems offer significant flexibility, robustness, and resistance to disturbances. Research into distributed cooperative control systems not only helps us better understand natural laws but also lays a solid theoretical foundation for the application of multi-agent systems in engineering practice. Currently, distributed cooperative control systems have been widely applied in various fields, including multi-vehicle/robot systems, communication networks, autonomous highway systems, and satellite clusters.

From an implementation perspective, utilizing multi-agent platforms (such as drones) to autonomously form swarms requires a high degree of cooperative ability. It can be said that cooperative control is the foundation of multi-agent swarm technology, and the collective intelligence that emerges from cooperative behavior is the core of multi-agent swarm technology [21]. However, for the cooperative control tasks of large-scale multi-agent systems, traditional centralized cooperative control methods struggle to address real-time processing issues arising from the system's scale. In stark contrast, nature demonstrates that millions of starlings can effortlessly perform graceful aerial dynamic formations, and billions of sardines can exhibit perfect swarm movement during migration. Therefore, distributed cooperative control technology, inspired by nature, is considered an effective solution for large-scale multi-agent system swarm cooperation [22, 23].

Take the recent surge in popularity of drone light shows for example, whether it was the 2015 Intel-created hundred-drone light show or the current Guinness World Record holder—Ehang's 1,374-drone light show in Xi'an in May 2018—centralized cooperative control technology was used. This means that, based on GPS navigation and positioning, each drone's fixed flight path was pre-designed, with no communication or interaction between the drones. Although Ehang's light show set a world record, the official performance the next day unexpectedly failed, revealing the shortcomings of centralized cooperative control methods when the system reaches a certain scale. Additionally, several teams, including those from Beihang University, Poisson Technology, and Roland University, have successfully conducted indoor and outdoor experiments with drone swarms based on distributed architectures, although the experimental scale remains within a few hundred drones. This suggests that distributed swarm cooperative technology for large-scale complex systems is still immature and not yet ready for stable application in practical projects. While large-scale swarm phenomena are common in nature, solving the swarm cooperative problem for large-scale multi-agent systems remains a significant challenge. Gaining a deeper understanding of the cooperative mechanisms behind swarm behavior and providing a reasonable explanation for the emergence of various swarm behaviors are crucial for advancing and improving distributed cooperative control technology.

3.3 Theoretical Tools

This section provides a brief introduction to some of the fundamental concepts, definitions, and lemmas that will be discussed in the following chapters, offering the necessary groundwork.

3.3.1 Graph Theory

Graph theory originated from the *Seven Bridges of Königsberg* problem. The problem questioned whether it was possible to start from any location and cross all seven bridges of the town without retracing any steps. Euler later reformulated this problem into a question of whether a graph could be drawn in a single stroke from the starting point to the endpoint, and ultimately concluded that there is no such path that could cross all seven bridges without repetition (Fig. 3.1). This study laid the foundation for the development of topology. By the mid-nineteenth century, many graph theory-related problems began to emerge, such as the famous Four Color Theorem and the Hamiltonian Circuit Problem [24].

The subject of study in Graph Theory is graphs. In Graph Theory, a graph is a structure consisting of a set of vertices and lines connecting pairs of vertices. Such a structure is typically used to describe specific relationships between certain entities. In practical applications, vertices represent the objects under study, while the lines between vertices indicate the relationships between the corresponding objects [25]. In multi-agent systems with a distributed architecture, the interaction topology is usually networked and can be clearly represented using graphs [26]. Therefore, algebraic graph theory becomes an essential analytical and modeling tool for studying distributed coordination control problems in multi-agent systems.

Here are some commonly used concepts in Graph Theory for reference.

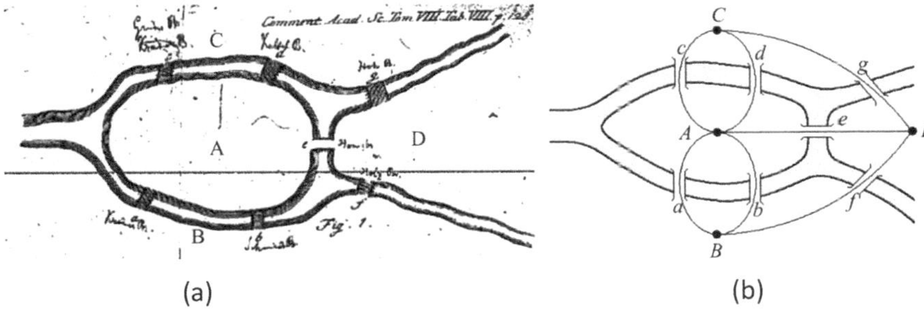

Fig. 3.1 Mathematical representation of the Seven Bridges problem **a** Euler's drawings **b** Mathematical diagram

1. Definition of a Graph

Let $G = (v, \varepsilon)$ represent a graph. A graph consists of a set of vertices v and a set of edges ε that represent the relationships between vertices. v represents a finite non-empty set of vertices, and ε represents a finite set of edges (relationships between vertices).

2. Undirected and Directed Graphs

A directed graph can be represented as (v, ε), where $v = \{1, 2, \ldots, N\}$ represents a finite non-empty set of vertices, and $\varepsilon \in v \times v$ is the set of ordered pairs of vertices, called edges. An edge (i, j) indicates that agent j can receive information from agent i, but not vice versa. Specifically, (i, i) is called a self-loop. If (i, j) is an edge in a directed graph, then i is defined as the parent vertex and j as the child vertex of this edge. Unlike directed graphs, in undirected graphs, the pairs of vertices representing edges are unordered, and an edge (i, j) indicates that entities i and j can exchange information with each other. An undirected graph can be seen as a special case of a directed graph, where an edge j in an undirected graph corresponds to both (i, j) and (j, i) in the directed graph.

3. Paths and Connectivity

A path in a directed graph is a finite sequence of vertices $i_1, i_2, \ldots, i_k$, such that $(i_s, i_{s+1}) \in \varepsilon$, $s = 1, 2, \ldots, k - 1$. If $i_1 = i_k$, then this path is called a cycle. If for any two distinct vertices i and j, there exists a path starting at i and ending at j, the directed graph is said to be strongly connected. A weak path in a directed graph is also a finite sequence of vertices $i_1, i_2, \ldots, i_k$, such that for any $s = 1, 2, \ldots, k - 1$, either $(i_s, i_{s+1}) \in \varepsilon$ or $(i_{s+1}, i_s) \in \varepsilon$, for $s = 1, 2, \ldots, k - 1$. If for any two distinct vertices i and j, there exists a weak path starting at i and ending at j, the directed graph is said to be weakly connected.

A weighted directed graph consists of a directed graph G and a non-negative matrix $C = [c_{ij}] \in R^{N \times N}$, where $(i, j) \in G \Leftrightarrow c_{ij} > 0$. The matrix C is called the weight matrix, and c_{ij} is the weight of the edge (j, i). If $(i, j) \in \varepsilon$ implies $(j, i) \in \varepsilon$, then G is called an undirected graph. If $C^T = C$, then $G(C)$ is a weighted undirected graph. An undirected graph can be viewed as a special type of directed graph; if an undirected graph is strongly connected, it is called connected.

4. Degree of a Vertex

In an undirected graph, the degree of a vertex ν is the number of edges connected to vertex ν, denoted as $D(\nu)$. In a directed graph, there are in-degrees and out-degrees. The in-degree refers to the number of edges terminating at the vertex, while the out-degree refers to the number of edges starting from the vertex. The degree of a directed graph equals the sum of the in-degree and out-degree of a vertex.

5. Balanced Graph

A directed graph G is called a balanced graph, if all nodes have equal in-degrees and out-degrees. All undirected graphs are balanced graphs.

6. Adjacency Matrix

The adjacency matrix of a graph $G = (\nu, \varepsilon)$ is represented by a non-negative matrix $A = \left[a_{ij}\right]_{N\times N}$, where

$$a_{ij} = \begin{cases} 1,\ (i,j) \in \varepsilon,\ i,j \in \nu \\ 0,\ \text{otherwise} \end{cases} \tag{3.1}$$

7. Degree Matrix

The degree matrix is a diagonal matrix $D = [d_{ij}]_{N\times N}$ where each diagonal element d_{ii} is the degree of vertex i, and off-diagonal elements are zero. It is constructed by summing the elements in each row of the adjacency matrix and placing these sums on the diagonal:

$$d_{ij} = \begin{cases} \deg(i),\ i = j \\ 0,\ otherwise \end{cases} \tag{3.2}$$

where $\deg(i) = \sum_j a_{ij}$。

8. Laplacian Matrix

The Laplacian matrix of a graph G, also known as the Kirchhoff matrix or admittance matrix, is commonly denoted as L. It is defined as $L = D - A$, where D is the degree matrix and A is the adjacency matrix of graph G.

3.3.2 *Consensus*

Let $x_i(t)$ represent the state information of individual i at time t. Consensus means that all individuals have consistent state information [28]. For example, when $t \to \infty$, we have $\left|x_i(t) - x_j(t)\right| \to 0, i, j = 1, \ldots, N,\ i \neq j$.

For instance, for a first-order continuous system, its consensus algorithm is as follows:

$$\dot{x}_i(t) = -\sum_{j=1}^{N} a_{ij}(t)(x_i(t) - x_j(t)) \tag{3.3}$$

3.3.3 Artificial Potential Field Method — Attractive and Repulsive Functions

The basic principle of the artificial potential field method is to view the environment in which an agent operates as a virtual force field. In this field, the target point exerts an attractive force on the agent, while the surrounding obstacles generate a repulsive force. The corresponding attractive and repulsive forces are obtained using the attractive potential field function and the repulsive potential field function.

In the case of fish schools, the attractive force is usually considered a long-range force based on visual perception, while the repulsive force is a short-range force based on the lateral line system of the fish (although it is generally stronger than the attractive force). Therefore, we consider the following potential functions:

$$f(p_{ij}) = -p_{ij}[f_a(\|p_{ij}\|) - f_r(\|p_{ij}\|)] \tag{3.4}$$

Here, f_a represents the magnitude of the long-range attractive force, while f_r represents the magnitude of the short-range repulsive force. $p_{ij} = p_i - p_j$ denotes the relative position vector from individual i to individual j, and $\|\cdot\|$ is the Euclidean norm. When the distance between individuals is δ, the attractive and repulsive forces balance each other, such as when $f_a(\delta) = f_r(\delta)$. Additionally, when $\|p_{ij}\| > \delta$, we have $f_a(\|p_{ij}\|) > f_r(\|p_{ij}\|)$; when $\|p_{ij}\| < \delta$, we get $f_a(\|p_{ij}\|) < f_r(\|p_{ij}\|)$. Note that the function $f(\cdot)$ is an odd function, such as $f(-p_{ij}) = -f(p_{ij})$.

There exist corresponding potential functions J_a and J_r satisfying

$$\nabla_{p_{ij}} J_a(\|p_{ij}\|) = p_{ij} f_a(\|p_{ij}\|) \tag{3.5}$$

and

$$\nabla_{p_{ij}} J_r(\|p_{ij}\|) = p_{ij} f_r(\|p_{ij}\|) \tag{3.6}$$

where $J_a(\|p_{ij}\|)$ and $J_r(\|p_{ij}\|)$ can be considered as the attractive and repulsive potentials, respectively. The potential energy between individuals i and j can be expressed as

$$V(\|p_{ij}\|) = J_a(\|p_{ij}\|) - J_r(\|p_{ij}\|) \tag{3.7}$$

The potential energy of individual i is

$$V_i = \sum_{j=1}^{N} V(\|p_{ij}\|) \tag{3.8}$$

whose gradient is

$$\nabla V_i = \sum_{j=1}^{N} \nabla_{p_{ij}} V(\|p_{ij}\|) \tag{3.9}$$

3.3.4 Lyapunov Stability Theorem

The stability of a system is a crucial characteristic during its operation. In industrial production, an unstable system carries significant uncertainty and danger. For instance, an unstable steel production system may experience severe explosions if disturbed by changes in environmental parameters such as oxygen content. An unstable facial recognition system may encounter system paralysis, misidentification, information leakage, and other issues when subjected to a sudden attack, posing threats to the user's personal safety, property, and information security. Similarly, an unstable temperature control system may cause sudden fluctuations in water temperature, affecting daily use and leading to problems such as skin burns [29]. Clearly, the stability of a system is the foundation for its normal operation, making the stability analysis of a system an important task after designing a control algorithm.

The stability criterion proposed by the famous Russian mathematician Aleksandr Lyapunov in 1892 is a widely used method for determining stability in the field of automatic control. Lyapunov stability describes a dynamical system's ability to keep its trajectories close to the equilibrium state under any initial conditions. Lyapunov stability theory is applicable to the stability analysis of linear systems, nonlinear systems, time-invariant systems, and time-varying systems. Lyapunov's first method, also known as the indirect method, involves linearizing the system's state equations and examining the distribution of eigenvalues of the linear system to determine its stability. Lyapunov's second method does not require solving the state equations and can address the stability determination of more complex nonlinear systems. However, selecting the Lyapunov function $V(x)$ generally requires skill and experience. The following sections will provide a detailed introduction to the Lyapunov stability criterion.

(1) Equilibrium Point and Equilibrium State

An important concept in Lyapunov stability theory is the equilibrium point and equilibrium state. The understanding of stability in a control system is based on the equilibrium point. Generally, the state equation of a system can be expressed as $\dot{x}(t) = f(x(t))$. The equilibrium point is the point at which the system's state does not change, mathematically corresponding to the point where the derivative is zero. Since the left side of the state equation is the derivative, setting the right side $\dot{x}(t) = f(x(t))$ and solving for x yields the system's equilibrium point. There may be more than one equilibrium point, and slight disturbances may render it unstable. In this case, the system's state at the equilibrium point is called the equilibrium state [30].

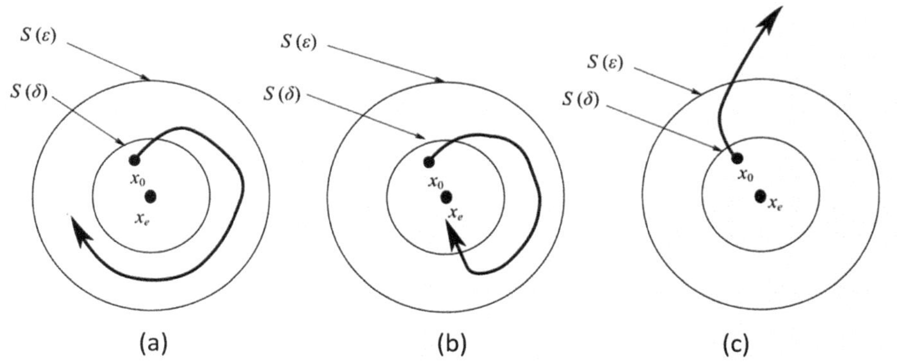

Fig. 3.2 The stability problem of Lyapunov equilibrium point **a** Lyapunov stabilization **b** asymptotic stability **c** instability

By analyzing the system's state after it is disturbed from the equilibrium state, equilibrium points can be classified into stable, asymptotically stable, globally asymptotically stable, and unstable categories. If the system remains near the equilibrium state x_e after a disturbance, x_e is said to be stable in the Lyapunov sense. If the system eventually converges to x_e, the equilibrium state is asymptotically stable in the Lyapunov sense. Global asymptotic stability differs in that it requires the system to eventually converge to x_e under any disturbance. Unstable means that the system's state begins to deviate from x_e after a disturbance and cannot return. This is illustrated in Fig. 3.2.

Combined with the diagram, stability in the Lyapunov sense is rigorously expressed as follows: For any $\epsilon > 0$, there exists a $\epsilon > 0$ such that if $\|x(0) - x_e\| < \delta$, then for all $t > 0$, we have $\|x(t) - x_e\| < \epsilon$. Here, $\|x(t) - x_e\|$ represents the deviation between the system state at time t and the equilibrium state, measured by the spatial distance between them. $x(0)$ denotes the initial state of the system.

(2) Lyapunov Stability Criteria

For a linear time-invariant system, the state equation can be written as $\dot{x}(t) = Ax(t)$. The solution to the state equation is the sum of the modes corresponding to the eigenvalues λ of matrix A. The system is stable if all eigenvalues λ lie in the left half of the complex plane, meaning all the real parts of λ are negative. Linear systems are relatively easy to calculate, so by linearizing a nonlinear system near its equilibrium state, the distribution of eigenvalues can be used to analyze the stability of the original nonlinear system. This is known as Lyapunov's first method, which determines stability by analyzing the distribution of characteristic roots, similar in concept to root locus, Nyquist stability criterion, and Routh criterion. In engineering practice, for complex nonlinear systems like those in aviation, it is common to perform appropriate linearization before conducting stability analysis for the entire system [31].

Lyapunov's second method is based on an idea analogous to a physical system. If the system's energy decreases monotonically over time, the system will eventually come to rest in a specific state. The second Lyapunov theorem does not require solving the state equations or linearizing the system, and is therefore called the direct method. It is described as follows.

Theorem 3.1 *If there exists a scalar function* $V(x)$ *such that* $V(x) = 0$ *if and only if* $x = 0$, *when* $x \neq 0$, *we have* $V(x) > 0$ *and* $\dot{V}(x) = \frac{d}{dt}V(x) = \sum_{i=1}^{n} \frac{\partial V}{\partial x_i} f_i(x) \leq 0$ *when* $x \neq 0$, *then the system is stable in the Lyapunov sense. Specifically, if* $\dot{V}(x) < 0$ *when* $x \neq 0$, *the system is asymptotically stable.*

Using the above theorem, the stability of the system can be easily determined under an appropriate Lyapunov function $V(x)$. The Lyapunov stability criterion can be applied to determine the stability of multi-agent systems.

3.4 Key Technologies

Inspired by swarm behavior, the concept of Multi-Agent Distributed Cooperative Control Systems (MADCCS) was proposed for multi-agent systems. MADCCS refer to complex systems that leverage a large number of low-cost, adaptable, and easily deployable unmanned devices to create a scale advantage. These systems accomplish designated cooperative tasks using local communication and simple rules. Compared to a single unmanned device, MADCCS provide a safer, more efficient, cost-effective, and flexible solution, making them highly promising for engineering applications.

The industry application of unmanned devices has been ongoing for years, primarily focusing on single-unit operations. However, collaborative task execution by swarms of unmanned devices is the inevitable direction for the future of the industry. This approach can truly meet the industry's needs for real-time, efficient, and accurate spatial information collection. At present, the interest in swarm technology within the civilian sector remains mostly academic. The implementation of collaborative swarms of unmanned devices involves high technical barriers, with five key technologies at the forefront:

(1) **Swarm Control Algorithms**: These ensure effective collaborative operation among unmanned devices.
(2) **Communication Technology**: During collaborative operations, each unmanned device acts as a communication network node to exchange information (e.g., location data). The communication network needs to allocate resources based on the task while ensuring high communication quality.
(3) **Coupling of Control Algorithms and Communication Technology**: This involves integrating the control algorithms and communication technology mentioned above. Under current communication quality constraints, the control algorithms must provide appropriate methods for controlling the unmanned

swarm, ensuring that the swarm meets both communication and collaborative task requirements.

(4) **Task Allocation Technology**: This technology enables unmanned devices to autonomously decompose complex tasks and respond based on their current status.

(5) **Obstacle Avoidance and Path Planning**: The algorithms used must ensure that unmanned devices can efficiently replan their paths in real-time when faced with unexpected situations and continuously adjust their routes based on threat information.

It is evident that swarm cooperative control technology is the core that allows unmanned devices to achieve a scale advantage. Technologies such as communication, sensing, and image recognition have developed in response to the needs arising from swarm cooperative control architecture. Therefore, this book focuses on proposing solutions for MADCCS from the perspective of swarm cooperative control technology. The emphasis on swarm strategy and swarm concepts is crucial, not only because swarms are typical distributed systems, but also because only through swarm strategies can multi-agent system cooperative control theory be directly scaled from small-scale to large-scale applications, laying a solid theoretical foundation for future engineering applications.

3.5 Summary

This chapter begins by introducing the concept of swarm systems and the essence of swarm intelligence from the perspective of multi-agent systems. It then provides a detailed explanation of the distributed cooperative control problem in multi-agent systems and the basic theoretical tools involved in solving this problem. Finally, the chapter outlines the key common technologies related to this issue, emphasizing that distributed cooperative control technology is the critical factor in achieving successful implementation.

References

1. Couzin, L. D., Krause, J., Franks, N. R., et al. (2005). Effective leadership and decision-making in animal groups on the move. *Nature, 433*(7025), 513–516.
2. Cucker, F., & Smale, S. (2007). Emergent behavior in flocks. *IEEE Transactions on Automatic Control, 52*(5): 852–862.
3. Czirok, A., Jacob, E. B., Cohen I., Vicsek, T. (1996). Formation of complex bacterial colonies via self-generated vortices. *Physics Review E, 54*(2), 1791–1801.
4. Bonabeau, E., Dorigo, M., Theraulaz, G. (1999). *Swarm intelligence: From natural to artificial systems*. Oxford University Press.
5. Minsky, M. (1986). *The society of mind*. Simon Schuster, Inc.

6. Hannebauer, M. (2002). *Autonomous dynamic reconfiguration in multiagent systems: Improving the quality and efficiency of collaborative problem solving*. Springer-Verlag.
7. Weiss, G. (1999). *Multiagent systems: A modern approach to distributed artificial intelligence*. MIT Press.
8. Gazi, V., Passino, K. M. (2003). Stability analysis of swarms. *IEEE Transactions on Automatic Control, 48*(4), 692–697.
9. 楚天广, 陈志福, 王龙, 等. 群体动力学与协调控制. 中国控制会议, 2007, 1, 611–614. (in Chinese)
10. Vicsek, T., Czirók, A., Benjacob, E., et al. (1995). Novel type of phase transition in a system of self-driven particles. *Physical Review Letters, 75*(6),1226.
11. Reynolds, C. W. (1987). *Flocks, herds, and schools: A distributed behavioral model*. St.Louis: ACM.
12. Shi, H., Wang, L., Chu, T. (2006). Virtual leader approach to coordinated control of multiple mobile agents with asymmetric interactions. *Physica D: Nonlinear Phenomena, 213*(1), 51–65.
13. Leonard, N. E., Fiorelli, E. (2001). Virtual leaders, artificial potentials and coordinated control of groups. In *Proceedings of the IEEE Conference on Decision and Control*, pp. 2968–2973.
14. Lee, D., Spong, M. W. (2007). Stable flocking of multiple inertial agents on balanced graphs. *IEEE Transactions on Automatic Control, 52*(8), 1469–1475.
15. Dimarogonas, D. V., Egerstedt, M, Kyriakopoulos, K.J. (2006). A leader-based containment control strategy for multiple unicycles. *IEEE Conference on Decision and Control*, 37(2), 1035-1042.
16. Borkar, V., Varaiya, P. P. (1982). Asymptotic agreement in distributed estimation. *IEEE Transactions on Automatic Control, 27*(3), 650–655.
17. Dimarogonas, D. V., Kyriakopoulos, K. J. (2008). A connection between formation infeasibility and velocity alignment in kinematic multi-agent systems. *Automatica, 44*(10), 2648–2654.
18. Ren, W. (2009). Collective motion from consensus with Cartesian coordinate coupling. *IEEE Transactions on Automatic Control, 54*(6), 1330–1335.
19. Smith, R. S., Hadaegh, F. Y. (2007). Closed-loop dynamics of cooperative vehicle formations with parallel estimators and communication. *IEEE Transactions on Automatic Control, 52*(8), 1404-1414.
20. Porfifiri, M., Roberson, D. G., Stilwell, D. J. (2007). Tracking and formation control of multiple autonomous agents: A two-level consensus approach. *Automatica, 43*, 1318–1328.
21. 韩亮, 任章, 董希旺, 李清东. 多无人机协同控制方法及应用研究. 导航定位与授时, 2018, *5*(4), 1–7. (in Chinese)
22. 贾永楠, 李擎. (2018). 多机器人编队控制研究进展. 系统工程学报, *8*, 893–900. (in Chinese)
23. Jia, Y. N., Li, Q., Zhang, W. C. (2019). A distributed cooperative approach for unmanned aerial vehicle flocking. Chaos, *29*(4), 043118.
24. 叶建川, 王江, 梁熠, 等. 四旋翼无人机前飞模态特性分析及建模. 兵工学报, 1–15[2022–02–24]. (in Chinese)
25. 秦文静, 林勇, 戚国庆. 基于一致性的无人机编队形成与防碰撞研究. 电子设计工程, 2018, *26*(09), 83–90. (in Chinese)
26. 王树禾. 图论. 北京: 科学出版社, 2009. (in Chinese)
27. 方保镕, 周继东, 李医民. 矩阵论. 北京: 清华大学出版社, 2013. (in Chinese)
28. 刘祖均, 何明, 马子玉, 等. 基于分布式一致性的无人机编队控制方法. 计算机工程与应用, 2020, *56*(23), 146–152. (in Chinese)
29. 卫强强, 刘蓉, 张衡. 四旋翼无人机高阶一致性编队控制方法. 电光与控制, 2019, *26*(8), 1–5. (in Chinese)
30. 宗群, 王丹丹, 邵士凯, 等. 多无人机协同编队飞行控制研究现状及发展. 哈尔滨工业大学学报, 2017, *49*(03), 1-14. (in Chinese)
31. 易文, 雷斌. 基于一致性的无人机编队飞行几何构型控制. 武汉科技大学学报, 2019, *42*(02), 150-154. (in Chinese)

Chapter 4
Application Scenarios and Countermeasures of Distributed Cooperative Control Systems with Egalitarian Agents

4.1 Typical Application Scenarios

4.1.1 Communication Networks

When a natural disaster occurs, the primary task is to establish a temporary communication network, assess the disaster situation, and then deploy helicopters to transport supplies and personnel. Besides expensive satellite communication methods, drone cooperation is the best solution to address communication issues in disaster relief.

The Intelligent Systems Laboratory at the École Polytechnique Fédérale de Lausanne (EPFL) developed a swarm of micro-drones for their micro-swarm aerial vehicle network project, which can rapidly set up communication networks in disaster-stricken areas (Fig. 4.1). These micro-drones can overcome terrain challenges and quickly deploy in disaster areas, using themselves as nodes to restore communication networks in the shortest possible time, providing a flexible solution for communication assurance in emergency rescue operations.

4.1.2 Water Quality Monitoring and Search

To study the collective behavior of micro underwater drones, the European Union established the CoCoRo project in 2011. CoCoRo stands for Collective Cognitive Robotics, with a funding of 2.9 million euros. Under this project, researchers at the Artificial Life Laboratory of the University of Graz in Austria released a swarm of 41 micro underwater robots in September 2015, which at that time was the largest underwater robot swarm in the world. The primary purpose of this swarm is for underwater monitoring and search operations. The system is scalable, reliable, and flexible in its behavioral capabilities. Researchers conducted experiments inspired by behavioral science and psychology to determine whether the robot network could

Y. Jia, *Distributed Coordination Control System*,
https://doi.org/10.1007/978-981-96-0365-7_4

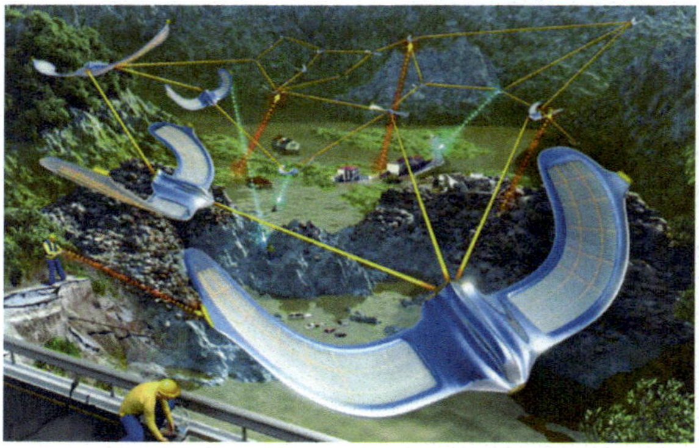

Fig. 4.1 Microswarm aircraft Network project at EPFL

exhibit collective cognition, comparing the swarm intelligence of this system with biological swarms in nature.

Each CoCoRo robot has a planar size of approximately 120 mm and is equipped with three actuators to control two degrees of freedom in the horizontal plane (translation or rotation) and one degree of freedom in the vertical plane (buoyancy). The robots communicate with each other using short-range LED light signals, and they are also equipped with RF communication and sonar for long-distance communication and positioning. Additionally, they have onboard optical sensors, inertial sensors, and pressure sensors.

4.1.3 Search and Emergency Rescue

Researchers at Loughborough University have developed a swarm drone system to assist in mountain search and rescue operations. This system consists of 10 small manually launched drones, each equipped with thermal imaging cameras, enabling them to easily locate missing hikers. The drones communicate with each other to ensure full coverage of the search area.

The Micro Aerial Vehicle Laboratory at Delft University of Technology is developing a swarm of pocket drones, each roughly the size of a palm. These drones can fly indoors and assist rescuers in searching for survivors in buildings that have suffered severe damage.

4.1.4 Smart Transportation

EHang, founded in 2014, is a high-tech company that integrates research, development, production, sales, and services for intelligent aircraft. The company has been recognized by international authoritative media as one of the "Global Best

Innovative Companies" and ranks among the top three drone companies worldwide. EHang offers a range of aircraft products and solutions, including autonomous aerial vehicles, smart city command and dispatch centers, industrial application-connected drones, automated drone swarms, and drone-based logistics delivery. In April 2019, EHang and Austria's largest aerospace manufacturer, FACC, jointly launched the "Urban Air Mobility" (UAM) project, aiming to solve urban traffic congestion by converting 2D transportation into 3D transportation through the coordinated use of sea, land, and air vehicles.

4.2 Countermeasure Research

The solutions discussed in this chapter are primarily aimed at multi-agent systems with the following characteristics: each agent possesses equal communication capabilities, equal positioning abilities, and a consistent decision-making mechanism. Such systems can also be referred to as swarm systems.

The primary goal of the solutions provided in this chapter is to achieve collective coordination tasks for the multi-agent system, where all agents move at the same speed, in the same direction, and maintain a compact formation. Throughout the process, each agent is capable of avoiding collisions with others, and the formation remains robust. Even if the formation is temporarily disrupted by external disturbances, it can quickly and automatically adjust back to a compact formation and maintain it.

This chapter focuses on robotic fish as the subject of study and provides specific solutions.

4.2.1 Introduction

The coordination control of multiple robotic fish has attracted increasing attention from scholars around the world, owing to its potential applications to kinds of underwater tasks such as environmental survey, search and rescue, ocean exploration, etc. In recent years, rapid advances in sensing and communication technologies have led to the possibility of the realization of the mature coordination control strategies of multi-agent systems on modern vehicles.

Flocking is one of the most significant research fields in coordination control of multi-agent systems, and flocking phenomena can be widely observed among fish in nature. In the colony, each agent depends on local sensing and simple rules to coordinate its behavior for the sake of keeping a common velocity while moving as a compact group. The classical flocking model was proposed by Reynolds [1], consisting of collision avoidance, velocity matching, and flocking centering. Since then, many variants of these three properties have been suggested, and relative control algorithms have also been proposed [2–9].

Therein, some authors have investigated the applications of these flocking algorithms to mobile vehicles. For example, Savkin et al. proposed a decentralized method for global flocking formation of multiple unicycles with velocity constraints, but they only involved the heading control of agents [10]. Gu et al. investigated a leader–follower flocking system consisting of a network of first-order unicycles with varying forward speeds and varying rotational speeds, and successfully applied the flocking algorithm to a group of real robots called "wifibots" [11]. Regmi et al. gave an experimental implementation of flocking algorithms in wheeled mobile robots, but didn't take the problem of finite communication into consideration [12]. Su et al. presented one solution to deal with the finite communication problem by adding constraint conditions to the potential function. Su et al. proposed a method to address the issue of limited communication by using potential function constraints [13]. However, using potential functions to form a formation often leads the system to fall into local equilibrium states [14]. Sabattini et al. addressed this by designing an appropriate potential function that ensures the system has a unique solution for the specified geometric formation [15]. Falconi et al. assumed that the communication topology of the system is a complete graph and designed a formation control strategy based on consensus using attraction/repulsion functions for agents with full dynamic attributes [16]. Due to the limitations of underwater communication technology, the implementation of control algorithms for underwater devices is restricted to surface communication or limited communication methods [17]. Furthermore, according to Weihs' dynamic analysis of fish schools, in the context of horizontal plane swarm control, the aggregated formation structure is an optimized configuration that maximizes energy efficiency [18]. This chapter focuses on the swarm control problem of a multi-robotic fish system under the constraint of limited communication.

4.2.2 Problem Formulation

Although the three-dimensional motion capabilities of robotic fish have been extensively studied [19], the execution of coordination algorithms for underwater devices is still limited to surface or wired communication due to constraints in underwater communication and positioning technologies. As a result, robotic fish, which generate propulsion through flexible bodies and tail fins, are often modeled as first-order unicycle models moving on the water surface [20, 21]. Given that the propulsion of a robotic fish mainly comes from the rear half of its body, we employ a second-order unicycle model, where the geometric center and the center of mass do not coincide, to further characterize the motion of the robotic fish. Let $\mathbb{N}$ represent the set of positive integers, and $\mathbb{R}$ represent the set of real numbers. Consider a multi-robotic fish system composed of N surface-swimming multi-joint bionic robotic fish.

As shown in Fig. 4.2, the kinematic characteristics of robotic fish i ($i = 1, \ldots, N$) can be described using the following extended second-order unicycle model:

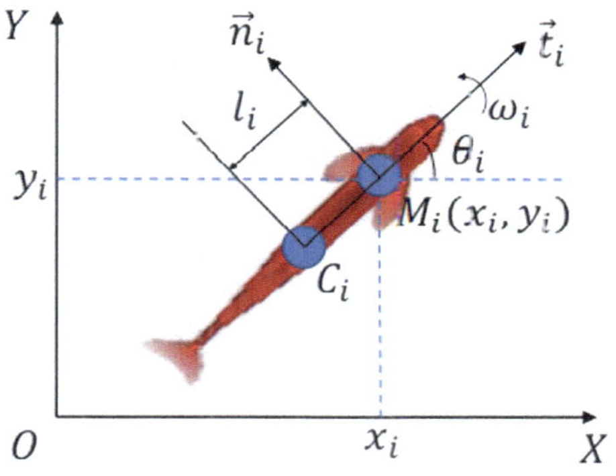

Fig. 4.2 The simplified model of robotic fish

$$
\begin{aligned}
\dot{x}_i(t) &= v_i(t)\cos\theta_i(t) - \omega_i(t) l_i \sin\theta_i(t) \\
\dot{y}_i(t) &= v_i(t)\sin\theta_i(t) + \omega_i(t) l_i \cos\theta_i(t) \\
\dot{\theta}_i(t) &= \omega_i(t) \\
\dot{v}_i(t) &= a_i(t) + \delta_i^1(t) \\
\dot{\omega}_i(t) &= b_i(t)/l_i + \delta_i^2(t)/l_i
\end{aligned}
\tag{4.1}
$$

where $P_i(t) = \left[x_i(t), y_i(t)\right]^T \in \mathbb{R}^2$ denotes the position vector of robotic fish i at time t, $\theta_i(t) \in R$ the heading angle of robotic fish i at time t, $v_i(t) \in \mathbb{R}$ the forward speed of robotic fish i at time t, $\omega_i(t) \in \mathbb{R}$ the rotational speed of robotic fish i at time t, $l_i > 0$ the distance between the geometrical center C_i and the mass center $M_i(x_i, y_i)$ of robotic fish i, $\vartheta_i(t) = \omega_i(t) l_i \in \mathbb{R}$ the tangential speed of robotic fish i at time t, $a_i(t) \in \mathbb{R}$ the forward acceleration of robotic fish i at time t, $b_i(t) \in \mathbb{R}$ the rotational acceleration of robotic fish i at time t. $\delta_i^1(t) \in \mathbb{R}$ the uncertain term added to the forward acceleration of robotic fish i at time t, $\delta_i^2(t) \in \mathbb{R}$ the uncertain term added to the rotational acceleration of robotic fish i at time t, $\delta_i(t) = \left[\delta_i^1(t), \delta_i^2(t)\right]^T \in R^2$ the uncertain intrinsic dynamics and the external disturbance (such as water-wave disturbance and other uncertain terms) of robotic fish i at time t. $\vec{t}_i(t) = \begin{bmatrix} \cos\theta_i(t) \\ \sin\theta_i(t) \end{bmatrix}$ and $\vec{n}_i(t) = \begin{bmatrix} -\sin\theta_i(t) \\ \cos\theta_i(t) \end{bmatrix}$ are two orthogonal unit vectors, the direction of $\vec{t}_i(t)$ aligns with the forward direction of robotic fish i. Therefore, $a_i(t)$ represents the component of the acceleration of robotic fish i in the direction of $\vec{t}_i(t)$, and $b_i(t)$ represents the component of the acceleration in the direction of $\vec{n}_i(t)$. Similarly, $\delta_i^1(t)$ is the component of $\delta_i(t)$ in the direction of $\vec{t}_i(t)$, and $\delta_i^2(t)$ is the component of $\delta_i(t)$ in the direction of $\vec{n}_i(t)$. The angle $\theta_i(t) \in [0, 2\pi)$ is measured counterclockwise as positive. Individual differences are not considered, so $I_i = I_d$ where $i = 1, 2, \ldots, N$. I_d is a constant.

Let $q_i(\mathrm{t}) = [v_i(t), \vartheta_i(t)]^T$. The kinematic equation (4.1) can be represented by the following matrix:

$$\begin{aligned}\dot{p}_i(t) &= H_i^T(t)q_i(t)\\ \dot{q}_i(t) &= u_i(t)\end{aligned} \tag{4.2}$$

where $H_i(t) = \left[\vec{t}_i(t), \vec{n}_i(t)\right]^T = \begin{bmatrix} \cos\theta_i(t) & \sin\theta_i(t) \\ -\sin\theta_i(t) & \cos\theta_i(t) \end{bmatrix}$, $u_i(t) = [a_i(t), b_i(t)]^T$. Besides, $u_i(t)$ is the control input of robotic fish i.

Each robotic fish has a limited interaction capability, and it can only interact with its neighbors. Let $N_i(t)$ denote the neighbor set of robotic fish i at time t. The neighbor set of robotic fish i at initial time $t_0 = 0$ is defined as:

$$N_i(0) = \{jp_i(0) - p_j(0) < D, j = 1, \ldots, N, j \neq i\} \tag{4.3}$$

where $\|\cdot\|$ is the Euclidean norm, and $D > 0$ is the interaction radius. Each robotic fish is able to obtain the state information of its neighbors via interaction network. Then the interaction network of the robotic fish system can be represented by an undirected graph G(t) with node set $v = \{1, 2, \ldots, N\}$ and edge set $\varepsilon(\mathrm{t}) \in v \times v$. Here, v is finite and nonempty. For $i, j \in v$, if i and j are neighbors at time t, one has $(i, j) \in \varepsilon(t)$. An undirected path is a sequence of unordered edges of the form $(i_1, i_2), (i_2, i_3), \ldots$, where $i_k \in v$. An undirected graph is connected if any two nodes can be connected by an undirected path.

In order to clarify the neighbor relationship, graph $G(t)$'s adjacency matrix $A(t)$ and Laplacian matrix $L(t)$ are respectively introduced. $A(t) = \left[\omega_{ij}(t)\right]_{N\times N}$ is defined as

$$\omega_{\mathrm{ij}}(t) = \begin{cases} 1, (j, i) \int \varepsilon(t) \\ 0, \text{otherwise} \end{cases} \tag{4.4}$$

and $L(t) = \left[\omega_{ij}\right]_{N\times N}$ is defined as

$$L_{ij}(t) = \begin{cases} -\omega_{ij}(t), i \neq j \\ \sum\limits_{k=1, k\neq i}^{N} \omega_{ik}(t), i = j \end{cases} \tag{4.5}$$

For an undirected graph, the Laplacian matrix is symmetric and positive semi-definite.

Given that the initial interaction network $G(0)$ is an undirected connected graph. In order to preserve the connectivity of the interaction network $G(t)$, the hysteresis adding new edges to the network is introduced [19], such that

(1) if $(i, j) \in \varepsilon\left(t^-\right)$ and $p_i(t) - p_j(t) < D$, then $(i, j) \in \varepsilon(t)$, for $t > 0$;
(2) if $(i, j) \notin \varepsilon\left(t^-\right)$ and $p_i(t) - p_j(t) < D - \zeta$, where $0 < \zeta < D$, then $(i, j) \in \varepsilon(t)$, for $t > 0$.

Suppose that the interaction network $G(t)$ switches at t_r, $r = 1, 2, \ldots$, $G(t)$ is a fixed graph in each nonempty, bounded, and contiguous time-interval $[t_{r-1}, t_r)$.

It is not easy to draw the characteristics of uncertain disturbances in underwater environment. In order to simplify the problem, the special characteristics of underwater environment are not considered into many theoretical analysis on the coordination control of multiple underwater vehicles [18, 20, 21]. However, in most cases, the influences of the water-wave and other uncertain terms on the underwater vehicles cannot be ignored. In this paper, $\delta_i(t)$ is supposed to satisfy the following constraint conditions:

$$\begin{aligned}\delta_i^1(t) &= k_1 \sum_{j \in N_i(t)} (1 - 2r_1(t))\big(v_i(t) - v_j(t)\big) + k_2 r_2(t) l_d \omega_i(t) \\ \delta_i^2(t) &= k_1 \sum_{j \in N_i(t)} (1 - 2r_1(t)) l_d \big(\omega_i(t) - \omega_j(t)\big) - k_2 r_2(t) v_i(t)\end{aligned} \quad (4.6)$$

where $k_1 \in [0, 1)$ and $k_2 \in [0, 1)$ are constants, and $r_1(t)$, $r_2(t)$ respectively denote a random number between [0, 1) generated at time t.

The flocking control problem discussed in this chapter primarily focuses on the multi-robot fish system (4.1). Each robotic fish independently determines its behavior based on the state information of its neighbors. The total potential energy of the system is given by:

$$v(t) = \sum_{i=1}^{N} V_i(t) = \sum_{i=1}^{N} \sum_{j \in N_i(t)} V\big(\|p_{ij}(t)\|\big) \quad (4.7)$$

where $V_i(t)$ denotes the potential energy of robotic fish i at time t, and the definition of $V\big(\|p_{ij}(t)\|\big)$ is given by

Definition 4.1 (Potential Function) Potential $V\big(\|p_{ij}(t)\|\big)$ is a differentiable, nonnegative, radially unbounded function of the distance $\big(\|p_{ij}(t)\|\big)$ between agent i and j, such that.

(1) $V\big(\|p_{ij}(t)\|\big) \to \infty$, as $\|p_{ij}(t)\| \to 0$;
(2) $V\big(\|p_{ij}(t)\|\big) \to \infty$, as $\|p_{ij}(t)\| \to D$;
(3) $V\big(\|p_{ij}(t)\|\big)$ attains its unique minimum when agents i and j are located at a desired distance.

Thus, we can formally describe the flocking control problem for the multi-robot fish system as follows:

Definition 4.2 (Flocking Control) Consider a robotic fish system (4.1) under interaction network $G(t)$. A control protocol $u_i(t) = [a_i(t), b_i(t)]^T$ $i = 1, \ldots, N$ is required to force the solution of the system gradually converge to a stable equilibrium state that satisfies the following conditions:

(1) **Collision avoidance**: collision between interaction agents are avoided, that is $p_i(t) \neq p_j(t), i, j \in v, i \neq j, t > 0$.

(2) **Velocity matching**: all of the pairwise forward speed differences, the pairwise rotational speed differences, and the pairwise heading angle differences asymptotically converge to zero, that is $\lim_{t\to\infty}|v_i(t) - v_j(t)| = 0$, $\lim_{t\to\infty}|\theta_i(t) - \theta_j(t)| = 0$, $\lim_{t\to\infty}|\omega_i(t) - \omega_j(t)| = 0$, where $i, j \in \nu$, $i \neq j$.

(3) **Cohesion Formation**: all individuals move as close as possible to their neighbors, leading the total potential energy V of the system to reach a maximum, where $\mathrm{d}V/\mathrm{d}t = 0$.

4.2.3 Distributed Coordination Control Algorithm Design and Stability Analysis

In nature, fish schools often exhibit a distributed coordination pattern. Typically, there is no leader within the group, and each individual independently decides its speed and direction. Biologists have found that the schooling behavior of fish is closely related to the attraction/repulsion forces between individuals [22]. The attraction force between fish is based on long-range visual cues, while the repulsion force is driven by short-range sensory input from the lateral line. The behavior of the fish in the group is determined by the combined effects of these attraction and repulsion forces. The unique distance at which the attraction and repulsion forces are equal is commonly referred to in biology as the "equilibrium distance".

Relying on their wide-angle vision and sensitive lateral line system, the speed and direction of individuals within a fish school are highly synchronized, meaning that each fish adjusts its speed and direction based on the behavior of its neighbors. Therefore, it is essential to establish a feedback mechanism based on the differences between individuals and their neighbors, such as a consensus algorithm. Consensus algorithms have been widely applied to solve alignment problems in multi-agent coordination control [23, 24]. Hence, in this section, we will design a distributed control protocol to address the flocking control problem for the multi-robotic fish system (4.1) by integrating the consensus algorithm with the artificial potential field method.

4.2.3.1 Algorithm Design

According to the above descriptions, the control protocol of robotic fish i consists of two components:

$$u_i(t) = \begin{bmatrix} a_i(t) \\ b_i(t) \end{bmatrix} = \alpha_i(t) + \beta_i(t) \tag{4.8}$$

where $\alpha_i(t)$ is the consensus term, and $\beta_i(t)$ is the gradient-based term. The first part $\alpha_i(t)$ aims at making the agents move with consistent velocities by consensus algorithms. The second part $\beta_i(t)$ mainly affects the relative positions between individuals in fish schooling through the negated gradients of the attractive/repulsive functions.

Let $p_{ij}(t) = p_i(t) - p_j(t)$. Assume that motion is synchronous and that no time delays. The control protocol for agent $i(i = 1, \ldots, N)$ is defined as

$$
\begin{aligned}
a_i(t) = &- \sum_{j \in N_i(t)} \left(v_i(t) - v_j(t)\right) - \sum_{j \in N_i(t)} \dot{p}_{ij}(t)^T \vec{t}_i(t) \\
&- \sum_{j \in N_i(t)} \nabla_{p_{ij}(t)} V\left(p_{ij}(t)\right)^T \vec{t}_i(t) \\
b_i(t) = &-l_d \sum_{j \in N_i(t)} \left(\theta_i(t) - \theta_j(t)\right) - l_d \sum_{j \in N_i(t)} \left(\omega_i(t) - \omega_j(t)\right) \\
&- \sum_{j \in N_i(t)} \dot{p}_{ij}(t)^T \vec{n}_i(t) - \sum_{j \in N_i(t)} \nabla_{p_{ij}(t)} V\left(p_{ij}(t)\right)^T \vec{n}_i(t)
\end{aligned}
\tag{4.9}
$$

4.2.3.2 Stability Analysis

Then, we are in a position to state our main results on the flocking problem of multiple agents (4.1).

Theorem 4.1 (Flocking) *Consider a system of N agents (4.1) under interaction network G(t). Each agent is steered by the control protocol (4.9). Let the initial interaction network G(0) be an undirected connected graph. Then*

(1) Collisions between interaction agents are avoided.
(2) The connectivity of the interaction network is preserved at all times.
(3) All of the pairwise forward speed differences, the pairwise rotational speed differences, and the pairwise heading angle differences asymptotically converge to zero.
(4) The system approaches a cohesive configuration that minimizes the total potential.

Proof Let $q_i^*(t) = [0, l_d \theta_i(t)]^T$, $q^*(t) = \left[q_1^*(t)^T, \ldots, q_N^*(t)^T\right]^T$, and $q(t) = \left[q_1(t)^T, \ldots, q_N(t)^T\right]^T$. According to (4.1) and (4.9), one gets

$$
\begin{aligned}
\dot{q}(t) = &-(L(t) \otimes I_2) q^*(t) - (L(t) \otimes I_2) q(t) \\
&- \sum_{i=1}^{N} H_i(t) \sum_{j \in N_i(t)} \left(\dot{p}_i(t) - \dot{p}_j(t)\right) \\
&- \sum_{i=1}^{N} H_i(t) \sum_{j \in N_i(t)} \nabla_{p_{ij}(t)} V\left(\left\|p_{ij}(t)\right\|\right) + \delta(t)
\end{aligned}
\tag{4.10}
$$

where I_2 denotes 2×2 identity matrix, $L(t)$ is the Laplacian matrix of the interaction network, and $\delta(t) = \left[\delta_1(t)^T, \ldots, \delta_N(t)^T\right]^T$.

Let $\tilde{p}(t) = \left[p_{11}(t)^T, \ldots, p_{1N}(t)^T, \ldots, p_{N1}(t)^T, \ldots, p_{NN}(t)^T\right]^T$, $\theta(t) = [\theta_1(t), \ldots, \theta_N(t)]^T$. Consider the following positive semi-definite energy function

$$E(\tilde{p}, q, \theta) = \frac{1}{2}V(t) + \frac{1}{2}q(t)^T q(t) + \frac{1}{2}l_d^2\theta(t)^T L(t)\theta(t) \tag{4.11}$$

Consider the system (4.1) with control input (4.9) on time interval $[t_{r-1}, t_r)$, where the interaction network of the multi-agent system is fixed. Then the derivative of $E(\tilde{p}, q, \theta)$ w.r.t. the time $t \in \left[t_{r-1}, t_r\right)$ is

$$\begin{aligned}
\frac{dE}{dt} &= \frac{1}{2}\sum_{i=1}^{N}\sum_{j\in N_i}\dot{p}_{ij}(t)^T\nabla_{p_{ij}(t)}V\left(\left\|p_{ij}(t)\right\|\right) \\
&\quad + q(t)^T\dot{q}(t) + = \sum_{i=1}^{N}\dot{p}_i(t)^T\sum_{j\int N_i}\nabla_{p_{ij}(t)}V\left(\left\|p_{ij}(t)\right\|\right) \\
&\quad + q(t)^T(-(L\otimes I_2)q^*(t) - (L\otimes I_2)q(t) \\
&\quad - \sum_{i=1}^{N}H_i(t)\sum_{j\int N_i}\left(\dot{p}_i(t) - \dot{p}_j(t)\right) - \sum_{i=1}^{N}H_i(t)\sum_{j\int N_i}\nabla_{p_{ij}(t)}V\left(\left\|p_{ij}(t)\right\|\right) \\
&\quad + \delta(t)) + l_d^2\dot{\theta}(t)^T L\theta(t) = \sum_{i=1}^{N}q_i(t)^T H_i(t)\sum_{j\int N_i}\nabla_{p_{ij}(t)}V\left(\left\|p_{ij}(t)\right\|\right) \\
&\quad - l_d^2\omega(t)^T L\theta(t) - q(t)^T(L\otimes I_2)q(t) \\
&\quad - \sum_{i=1}^{N}q_i(t)^T H_i(t)\sum_{j\int N_i}\left(\dot{p}_i(t) - \dot{p}_j(t)\right) \\
&\quad - \sum_{i=1}^{N}q_i(t)^T H_i(t)\sum_{j\int N_i}\nabla_{p_{ij}(t)}V\left(\left\|p_{ij}(t)\right\|\right) \\
&\quad + \sum_{i=1}^{N}q_i(t)^T\delta_i(t) + l_d^2\omega(t)^T L\theta(t) \\
&= -q(t)^T(L\otimes I_2)q(t) - \sum_{i=1}^{N}\sum_{j\in N_i}\dot{p}_i(t)^T\left(\dot{p}_i(t) - \dot{p}_j(t)\right) \\
&\quad + k_1(1 - r_1(t))\sum_{i=1}^{N}\sum_{j\int N_i}q_i^T\left(q_i - q_j\right)
\end{aligned}$$

$$+ k_2 r_2(t) \sum_{i=1}^{N} \sum_{j \int N_i} (v_i(t) l_d \omega_i(t) - l_d \omega_i(t) v_i(t))$$
$$= -(1 - k_1(1 - 2r_1(t))(L \otimes I_2) q(t) - \dot{p}(t)^T (L \otimes I_2) \dot{p}(t) \leq 0 \quad (4.12)$$

where $p(t) = \left[p_1(t)^T, \ldots, p_N(t)^T\right]^T$, $\in (-1, 1)$.

The initial value of the total potential of the system is finite, and the initial speeds and the initial heading angles of all agents are also finite. Thus, the system's initial energy $E(\tilde{p}(0), q(0), \theta(0))$ is finite, and the supremum of E is obviously its initial value $E(\tilde{p}(0), q(0), \theta(0))$. Then, the total potential V is finite and hence, the potential $V\left(\left\|p_{ij}(t)\right\|\right)$ between agents i and j is finite.

Based on the nearest neighbor interaction rules, collisions may happen only between interaction agents. Suppose that there exists a time instant t^* when agent i and j collide with each other, namely, $\left\|p_{ij}(t)\right\| \to 0$ as $t \to t^*$. Then, according to rule (1) of Definition 4.1, one gets $V\left(\left\|p_{ij}(t)\right\|\right) \to \infty$ as $t \to t^*$. However, $V\left(\left\|p_{ij}(t)\right\|\right) \to \infty$ violates the conclusion that $V\left(\left\|p_{ij}(t)\right\|\right)$ remains finite. Thus, t^* cannot exist, that is to say, there is no collision between interaction agents during the whole process. Conclusion 1 of Theorem 5.1 is proved.

According to rule (2) of Definition 4.1, if for some $(i, j) \in \varepsilon$, $\left\|p_{ij}(t)\right\| \to D$, then $V\left(\left\|p_{ij}(t)\right\|\right) \to \infty$, which violates the conclusion that $V\left(\left\|p_{ij}(t)\right\|\right)$ remains finite. It follows that $\left\|p_{ij}(t)\right\|$ for all $(i, j) \in$ and $t \in [t_r, t_{r+1})$. Hence, whenever there is a link between two agents, the link is never lost during each time-interval $[t_r, t_{r+1})$, that is, $\varepsilon(t_r) \subseteq \varepsilon(t_{r+1})$. Note that if an edge $(i, j) \notin \varepsilon\left(t^-\right)$ is added to $\varepsilon(t)$, one gets $\left\|p_i(t) - p_j(t)\right\| < D - \zeta$, which ensures that, the associated potential $V\left(\left\|p_{ij}(t)\right\|\right)$ is bounded and hence, so is the new potential $V(t)$. Consequently, the connectivity of the communication network can be preserved at all times. Conclusion 2 of Theorem 4.1 is proved.

Assume that there are $m_r \in \mathbb{N}$ new links being added to the interaction network at switching time $t_r (r = 1, 2, \ldots)$. Since the initial interaction network $G(0)$ is a connected graph, the proposed control protocol can guarantee that the sequence of switching topologies $G(t_r)$ within $\left[t_r, t_{r+1}\right)$ consists of connected graphs, that is, $G(t_r) \in G_c$, where G_c denotes the set of all connected graphs on the vertices. The number of the vertices is finite, thus G_c is a finite set. Assume that there are at most M $\in$ N new links that can be added to the initial interaction network $G(0)$. Clearly, $0 < m_r \leq M$ and $r \leq M$. Therefore, the number of switching times of the system is finite, and the interaction network $G(t)$ eventually becomes fixed. Suppose that the last switching time is t_f, the following discussions are restricted on the time interval $\left[t_f, \infty\right)$.

Note that the distance between neighbors is not longer than D. Hence, the set

$$B = \{(\tilde{p}, q, \theta) | E(\tilde{p}, q, \theta) \leq E(\tilde{p}(0), q(0), \theta(0))\} \quad (4.13)$$

Is positively invariant, where$\tilde{p} \in D_p$, $D_p = \left\{\tilde{p} | \|p_{ij}\| \in (0, D), \forall (i, j) \in \varepsilon\right\}$. Since $G(t)$ is connected for all $t \geq 0$, one gets $\left\|p_{ij}\right\| < (N - 1)D$ for all i and j. Since

$E(\tilde{p}(t), q(t), \theta(t)) \leq E(\tilde{p}(0), q(0), \theta(0))$, one gets $q^T q \leq 2E(\tilde{p}(0), q(0), \theta(0))$, that is $\|q\| \leq \sqrt{2E(\tilde{p}(0), q(0), \theta(0))}$. In addition, $\theta_i \in [0, 2\pi), i = 1, \ldots, N$. Thus, the set B is closed and bounded, hence compact. Note that the system (4.1) with control input (4.9) is an autonomous system on the concerned time interval $\left[t_f, \infty\right)$. Then, according to the LaSalle-Krasovskii invariance principle [25], the trajectories of all agents will converge to the invariant set $S = \{(\tilde{p}, q, \theta)|\mathrm{d}E/\mathrm{d}t = 0\}$. Clearly, $dE/dt = 0$ if and only if $q_i = q_j$ and $\dot{p}_i = \dot{p}_j$. To this end, one gets $\lim\limits_{t\to\infty}\left|v_i(t) - v_j(t)\right| = 0$, $\lim\limits_{t\to\infty}\left|\omega_i(t) - \omega_j(t)\right| = 0$, and $\lim\limits_{t\to\infty}\left|\dot{p}_i(t) - \dot{p}_j(t)\right| = 0$.

According to (4.1), $\dot{p}_i(t) = \dot{p}_j(t)$ is equivalent to

$$\begin{cases} v_i(t)\cos\theta_i(t) - \omega_i(t)l_d \sin\theta_i(t) = v_j(t)\cos\theta_j(t) - \omega_j(t)l_d \sin\theta_j(t) \\ v_i(t)\sin\theta_i(t) + \omega_i(t)l_d \cos\theta_i(t) = v_j(t)\sin\theta_j(t) + \omega_j(t)l_d \cos\theta_j(t) \end{cases} \tag{4.14}$$

In steady state, one gets $q_i(t) = q_j(t)$, namely, $v_i(t) = v_j(t)$ and $\omega_i(t) = \omega_j(t)$. Without loss of generality, let $v_i(t) = v_d$ and $\omega_i(t) = \omega_d$ in steady state, then (4.14) can be rewritten as

$$\begin{cases} \left(v_d^2 + I_d^2\omega_d^2\right)\sin\theta_i(t) = \left(v_d^2 + I_d^2\omega_d^2\right)\sin\theta_j(t) \\ \left(v_d^2 + I_d^2\omega_d^2\right)\cos\theta_i(t) = \left(v_d^2 + I_d^2\omega_d^2\right)\cos\theta_j(t) \end{cases} \tag{4.15}$$

Robotic fish cannot keep stationary in the water due to uncertain disturbance ($v_d \neq 0$), thus one gets $\theta_i(t) = \theta_j(t) \in [0, 2\pi)$ in steady state. That is to say, $\lim\limits_{n\to\infty}\left(\theta_i(t) - \theta_j(t)\right) = 0$. Similarly, let $\theta_i(t) = \theta_d$ in steady state, then one gets $H_d = \begin{bmatrix} \cos\theta_d & \sin\theta_d \\ -\sin\theta_d & \cos\theta_d \end{bmatrix}$. Conclusion 3 of Theorem 4.1 is proved.

Since $\nabla_{p_{ij}(t)}V\left(\left\|p_{ij}(t)\right\|\right)$ is an odd function, one gets $\sum\limits_{i=1}^{N}\sum\limits_{j\in N_i}\nabla_{p_{ij}(t)}V\left\|\left(p_{ij}(t)\right)\right\|=0$. In steady state, note that $q_i^*(t) = q_j^*(t)$, $q_i(t) = q_j(t)$, and $\dot{p}_i(t) = \dot{p}_j(t)$, from the control input (4.9), one knows

$$\begin{aligned} \dot{\overline{q}}(t) &= \frac{1}{N}\sum_{i=1}^{N}\dot{q}_i(t) \\ &= \frac{1}{N}\left(-\sum_{i=1}^{N}\sum_{j\int N_i}\left(q_i^*(t) - q_j^*(t)\right) - \sum_{i=1}^{N}\sum_{j\int N_i}\left(q_i(t) - q_j(t)\right)\right) \\ &\quad - \sum_{i=1}^{N}H_i(t)\sum_{j\int N_i}\left(\dot{p}_i(t) - \dot{p}_j(t)\right) \\ &\quad - \sum_{i=1}^{N}H_i(t)\sum_{j\int N_i}\nabla_{p_{ij}(t)}V\left(\left\|p_{ij}(t)\right\|\right) \end{aligned}$$

$$= \frac{1}{N}\left(-\sum_{i=1}^{N} H_d \sum_{j \int N_i} \nabla_{p_{ij}(t)} V\left(\left\|p_{ij}(t)\right\|\right)\right)$$

$$= \frac{1}{N} H_d \left(-\sum_{i=1}^{N} \sum_{j \int N_i} \nabla_{p_{ij}(t)} V\left(\left\|p_{ij}(t)\right\|\right)\right) = 0 \tag{4.16}$$

which implies that the forward speed and the rotational speed of the center of mass of all agents are both invariant. Due to $q_1 = \ldots = q_N = \overline{q}$, one gets $\dot{q}_i = \dot{\overline{q}} = 0$, that is, $\dot{v}_i = 0$ and $\dot{\omega}_i = 0$, which implies that v_d and ω_d are constants. Thus, from (4.9), one obtains

$$\begin{cases} -\sum_{j \in N_i} \nabla_{p_{ij}(t)} V\left(\left\|p_{ij}(t)\right\|\right)^T \vec{t}_i(t) = 0 \\ -\sum_{j \in N_i} \nabla_{p_{ij}(t)} V\left(\left\|p_{ij}(t)\right\|\right)^T \vec{n}_i(t) = 0 \end{cases} \tag{4.17}$$

Since $\vec{t}_i(t) = \begin{bmatrix} \cos\theta_i(t) \\ \sin\theta_i(t) \end{bmatrix}$ and $\vec{n}_i(t) = \begin{bmatrix} -\sin\theta_i(t) \\ \cos\theta_i(t) \end{bmatrix}$ are unit vectors orthogonal to each other, as well as$I_d \neq 0$, (4.17) is equivalent to $\sum_{j \int N_i} \nabla_{p_{ij}(t)} V\left(\left\|p_{ij}(t)\right\|\right) = 0$. That is to say, the total potential reaches the minimum value due to

$$\frac{dV}{dt} = \sum_{i=1}^{N} \dot{p}_{ij}(t)^T \sum_{j \in N_i} \nabla_{p_{ij}(t)} V\left(p_{ij}(t)\right) = 2\sum_{i=1}^{N} \dot{p}_i(t)^T \sum_{j \in N_i} \nabla_{p_{ij}(t)} V\left(\left\|p_{ij}(t)\right\|\right) = 0 \tag{4.18}$$

Conclusion 4 of Theorem 4.1 is proved.

4.2.3.3 Simulation Results

Ten agents are assigned to validate the proposed theoretical results on Matlab. The initial interaction network of the ten agents is a connected graph. The interaction radius D = 30 m. The green star point is the initial position of each agent, while the center of the color ball is the final position of each agent at time t = 60 s. The simulation results are presented in Fig. 4.3. Figure 4.3a tells that all agents asymptotically converge to the desired formation with a cohesive configuration. Figure 4.3b–d respectively show that all agents asymptotically swim with the consistent heading angle, the consistent forward speed, and the consistent rotational speed. It should be noted that the value of the heading angle θi is exactly the remainder by dividing 2π, for we have clarified in Sect. 4.2.2 that the heading angle is limited to [0, 2π). Figure 4.3e gives the distances between any two agents. The distances are all asymptotically stable and there is no zero-value distance at all times. Thus, collisions between any two agents are totally avoided during the evolution process. Figure 4.3f provides the time response of the total potential of the ten-agent system in time-interval $\left[t_f, \infty\right)$, where $t_f = 0.67$ s. It is obvious that the total potential converges to a minimum value. Besides, the initial adjacency matrix of the system is

$$A(0) = \begin{bmatrix} 0&0&1&1&0&1&1&1&1&1 \\ 0&0&0&0&0&1&0&0&1&1 \\ 1&0&0&1&0&1&1&1&1&1 \\ 1&0&1&0&1&1&1&1&1&1 \\ 0&0&0&1&0&0&0&0&0&0 \\ 1&1&1&1&0&0&1&1&1&1 \\ 1&0&1&1&0&1&0&1&0&0 \\ 1&0&1&1&0&1&1&0&1&1 \\ 1&1&1&1&0&1&0&1&0&1 \\ 1&1&1&1&0&1&0&1&1&0 \end{bmatrix} \tag{4.19}$$

which indicates that the interaction network is a connected graph but not a complete graph. The connectivity preservation of the system can be concluded according to the adjacency matrixes of the system output in real time. Finally, the interaction network of the system is fixed, which can be expressed by the following adjacency matrix

$$A(60) = \begin{bmatrix} 0&1&1&1&1&1&1&1&1&1 \\ 1&0&1&1&1&1&1&1&1&1 \\ 1&1&0&1&1&1&1&1&1&1 \\ 1&1&1&0&1&1&1&1&1&1 \\ 1&1&1&1&0&1&1&1&1&1 \\ 1&1&1&1&1&0&1&1&1&1 \\ 1&1&1&1&1&1&0&1&1&1 \\ 1&1&1&1&1&1&1&0&1&1 \\ 1&1&1&1&1&1&1&1&0&1 \\ 1&1&1&1&1&1&1&1&1&0 \end{bmatrix} \tag{4.20}$$

Specially, during the simulations and the experiments, it is supposed that the flocking task is realized if only the error is no more than one percent. Thus, according to the numerical simulation results, the desired flocking phenomenon firstly emerges at about 10 s. Besides, ten agents maintain the flocking behavior from 10 to 60 s.

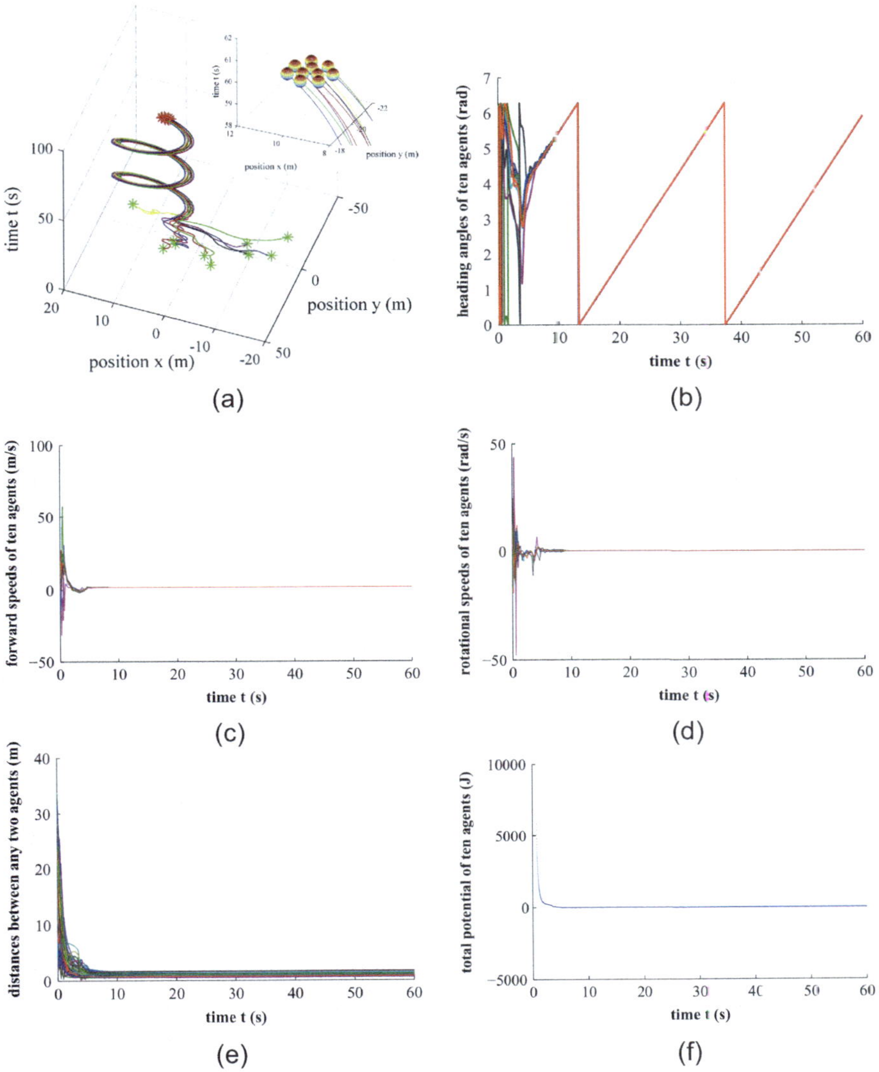

Fig. 4.3 Simulation results of flocking with switching interaction network. **a** trajectories of ten agents versus time. **b** heading angles of ten agents versus time. **c** forward speeds of ten agents versus time. **d** rotational speeds of ten agents versus time. **e** distances between any two agents versus time. **f** total potential of ten agents versus time

4.2.4 Experiment Validation and Discussion

4.2.4.1 Robotic Fish

The robotic fish under consideration is shown in Fig. 4.4. The size of the robotic fish is about 0.45 m × 0.05 m × 0.12 m (length × width × height). A communication module, a control module, and a power supply module are equipped inside the fish head. For the sake of imitating the undulation motion of real fish, the body of the robotic fish is constructed by connecting multiple joints in series. Information between robotic fish can be exchanged via the communication module.

The robotic fish takes carangiform movements [26], and its undulatory motion employs the swimming form of a traveling wave described by [27]

$$y_{body}(x,t) = A(x)\sin(kx + \omega t) \tag{4.21}$$

where y_{body} is the transverse displacement of the fish body, x is the displacement along the main axis, t is time, $A(x) = c_1 x + c_2 x^2$ is the wave amplitude,

c_1 and c_2 are adjustable parameters, $k = 2\pi/\lambda$ is the body wave number, λ is the body wave length, $\omega = \frac{2\pi}{M}$ is the circular frequency of oscillation, M is the frequency of oscillation.

To facilitate the physical implementation of the proposed traveling wave, each joint $\psi_i^j(t)(j = 1, 2, 3)$ of the robotic fish is controlled by

$$\psi_i^j(t) = \Phi_i^j(t) + A_i^j(t)\,_{\sin(2)}\pi f_i^j(t)t + \varphi_i^j) \tag{4.22}$$

where subscript j denotes the joint number, $\psi_i^j(t)$ denotes the angular position of the jth joint of robotic fish i at time t, $\Phi_i^j(t)$ denotes the angular offset of the jth joint of robotic fish i at time t, $A_i^j(t)$ denotes the amplitude of the jth joint of robotic fish i at time t, $f_i^j(t)$ denotes the oscillatory frequency of the jth joint of robotic fish i

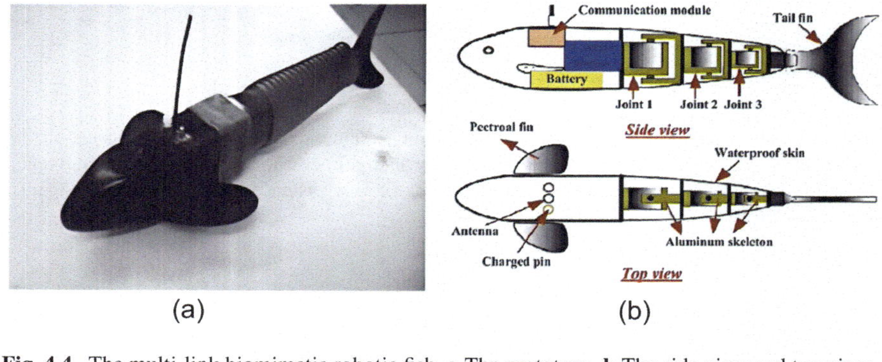

(a) (b)

Fig. 4.4 The multi-link biomimetic robotic fish. **a** The prototype. **b** The side view and top view

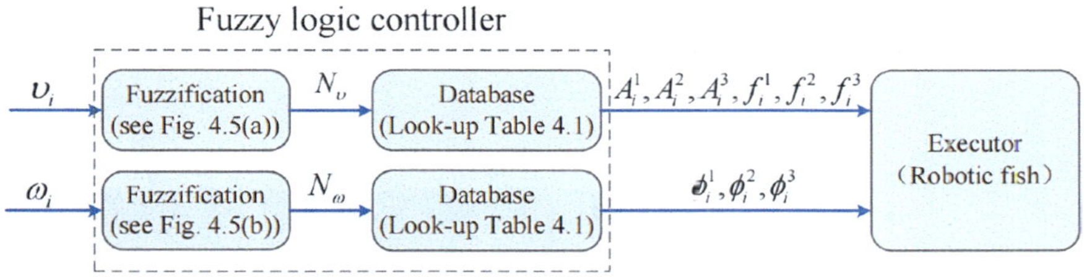

Fig. 4.5 The schematic diagram of the fuzzy logic controller

at time t, and φ_i^j denotes the lag angle of the jth joint of robotic fish i at time t. In particular, the lag angles of robotic fish i satisfy the constraints that $\varphi_i^2 - \varphi_i^1 = 40°$ and $\varphi_i^3 - \varphi_i^1 = 169.3°$ for $i = 1, \ldots, N$.

The motion of robotic fish i at time t is decided by these control commands $f_i^j(t)$, $A_i^j(t)$, and $\Phi_i^j(t)$ for $j = 1, 2, 3$. Therein, the forward speed $v_i(t)$ depends on the oscillatory frequency $f_i^j(t)$ and the amplitude $A_i^j(t)$, $j = 1, 2, 3$. The rotational speed $\omega_i(t)$ is mainly related to the angular offset $\Phi_i^j(t)$, $j = 1, 2, 3$. In order to build the relationship between the control commands A_i^j, f_i^j, Φ_i^j and the speeds ω_i, v_i, a fuzzy logic controller has been embedded into the bottom control of each robotic fish as shown in Fig. 4.5.

4.2.4.2 Experiment Platform

Figure 4.6 gives the experiment platform, which comprises an overhead camera, a host personal computer (PC), a RF communication system, and a swimming pool with dimension of 3 m × 2 m × 0.05 m (length × width × depth). Robotic fish serve as actuators. These robotic fish are not equipped with any sensors, and have limited computation/interaction abilities. Thus, the overhead camera is used to identify each robotic fish, and the host PC is used to compute the position and the heading angle of each robotic fish. Besides, robotic fish communicate with each other via the host PC as a base station to form the interaction network. The flocking algorithm is embedded into the base station to generate discrete control commands for the robotic fish.

4.2.4.3 Implementation Steps

The initial states of the robotic fish are randomly chosen. Let h be the sample period of the system. Then, the sampling instant can be expressed by $t = hk$, $k = 0, 1, 2, \ldots$. The realization process of the proposed algorithm for multiple robotic fish is given as follows. Firstly, (4.9) is used to compute the forward acceleration $a_i(t)$ and rotational acceleration $b_i(t)$ of robotic fish i at current sampling instant $t = hk$. Secondly, the forward speed $v_i(t)$ and the rotational speed $\omega_i(t)$ of robotic fish i at next sampling

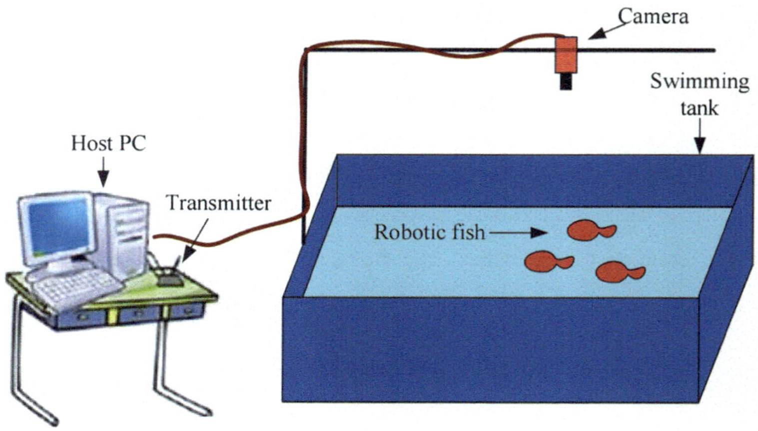

Fig. 4.6 Experiment platform for the flocking task of three robotic fish

instant $hk + h$ are respectively computed by

$$\begin{aligned} v_i(hk + h) &= v_i(hk) + a_i(hk) * h \\ \omega_i(hk + h) &= \omega_i(hk) + b_i(hk) * h \end{aligned} \tag{4.23}$$

Thirdly, according to the membership function shown in Fig. 4.7, the forward speed grade N_v and the rotational speed grade N_ω are respectively obtained. Finally, look up Table 4.1 to find out the relative control parameters A_i^j, f_i^j, Φ_i^j of each joint j of the robotic fish i at next sampling instant. The evolution process ends when the running time is larger than the preappointed time.

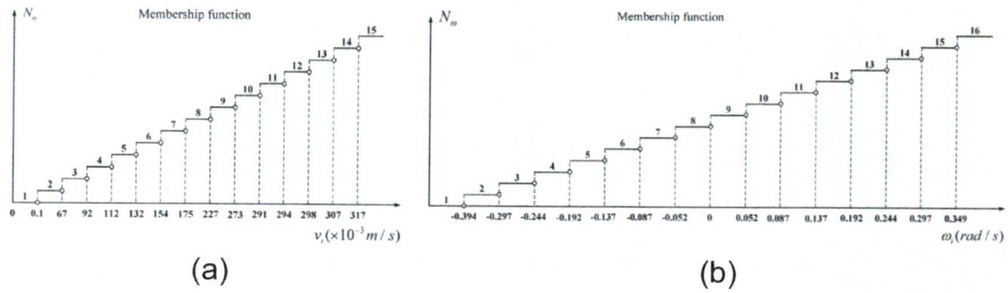

Fig. 4.7 The membership functions. **a** forward speed grade N_v versus forward speed $v_i(t)$. **b** rotational speed grade N_ω versus rotational speed $\omega_i(t)$

Table 4.1 Speed grades N_v, N_ω of the robotic fish i versus relative control commands A_i^j, f_i^j, Φ_i^j, $j = 1, 2, 3$

N_v	$A_i^1(°)$	$A_i^2(°)$	$A_i^3(°)$	$f_i^1 = f_i^2 = f_i^3 (Hz)$	N_w	$\phi_i^1(°)$	$\phi_i^2(°)$	$\phi_i^3(°)$
0	0	0	0	0	1	– 54	– 59	– 64
1	2	12	15	0.15	2	– 48	– 53	– 58
2	3	13	18	0.25	3	– 42	– 47	– 52
3	4	14	19	0.266	4	– 30	– 35	– 40
4	5	15	20	0.286	5	– 18	– 23	– 28
5	6	16	21	0.308	6	– 12	– 17	– 22
6	7	17	22	0.354	7	– 6	– 11	– 16
7	8	18	23	0.364	8	0	0	0
8	9	19	24	0.4	9	0	0	0
9	10	20	25	0.44	10	6	11	16
10	11	21	26	0.5	11	12	17	22
11	12	22	27	0.57	12	18	23	28
12	13	23	28	0.666	13	30	35	40
13	14	24	29	0.8	14	42	47	52
14	15	25	30	1	15	48	53	58
15	16	26	31	1.34	16	54	59	64

4.2.4.4 Simulation Results

As an auxiliary tool for the research on underwater equipment collaboration, the robotic fish simulation platform aims to advance the development of underwater devices and related technologies. This platform has been adopted as the standard platform for the 2D simulation category of underwater robots in the China Robot Competition. The platform is built on Microsoft's Robotics Developer Studio (MRDS) and is designed based on the kinematic characteristics and control mechanisms of robotic fish. Additionally, to simulate real-world environments, the platform takes into account external disturbances and the effects of boundary conditions.

Ten robotic fish are designated for the simulation validation. The simulation period h of the animation platform is about $h = 0.11$ s. Ren et al. have given the details on the boundary collision model, causing by the limited size of the swimming field [28]. Figure 4.8 gives successive simulation time snapshots of flocking. It is obvious that ten robotic fish quickly get close to each other meanwhile adjust their heading angles and speeds. Finally, they keep a cohesive flocking formation and swim with the same heading angle, the same forward speed, and the same rotational speed. Ten robotic fish do the uniform curve motion together.

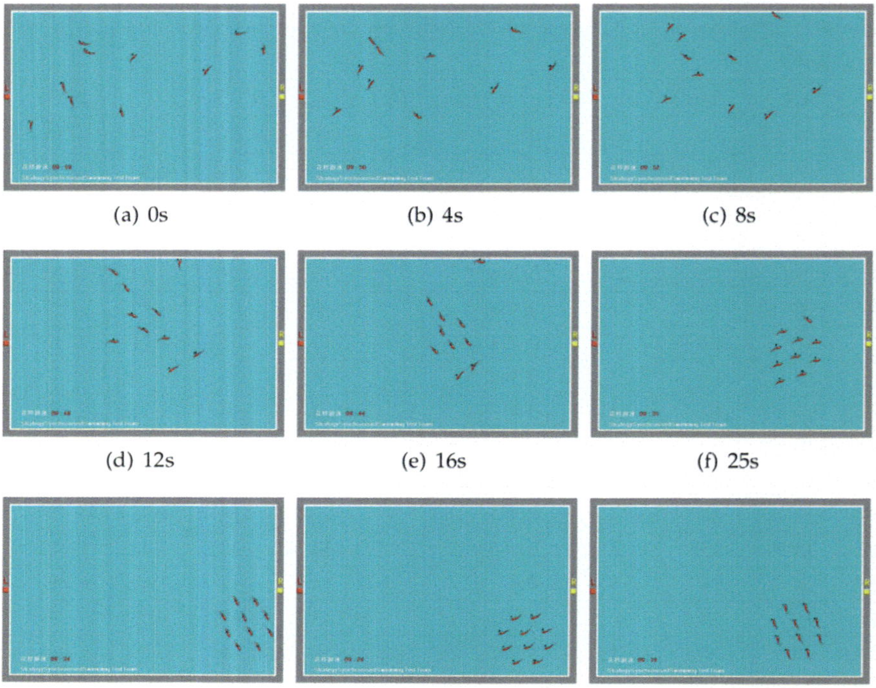

(a) 0s (b) 4s (c) 8s

(d) 12s (e) 16s (f) 25s

(g) 31s (h) 36s (i) 41s

Fig. 4.8 Successive simulation time snapshots of flocking of ten robotic fish

4.2.4.5 Experiment Results

The experiment aims to validate the practicality of the control protocol (4.9) for multiple robotic fish swimming in the water surface of the pool. Figure 4.9 shows the evolutionary process of the flocking experiment implemented by three robotic fish with a loose initial configuration and miscellaneous initial states. Three robotic fish show the desired cohesive flocking behavior at time 14 s. After that, fish 3 collides with the boundary of the swimming pool, and the triangle formation is temporarily destroyed. However, the flocking algorithm still works well to enable three robotic fish to recover the desired formation as soon as possible. It is obvious that the sustained flocking behavior of three robotic fish can be seen with the existence of field boundary.

Figure 4.10a gives the time responses of the trajectories of three robotic fish. The green circles denote their initial positions, and the red squares denote their final positions at time $t = 25$ s. Let $L = \begin{bmatrix} 0 & 1 & 1 \\ 1 & 0 & 1 \\ 1 & 1 & 0 \end{bmatrix}$ and $\theta = [\theta_1, \theta_2, \theta_3]^T$. Define that $\Theta = \frac{1}{2}\theta(t)^T L(t)\theta(t)$ is the energy function of the heading angle. Θ is used to evaluate

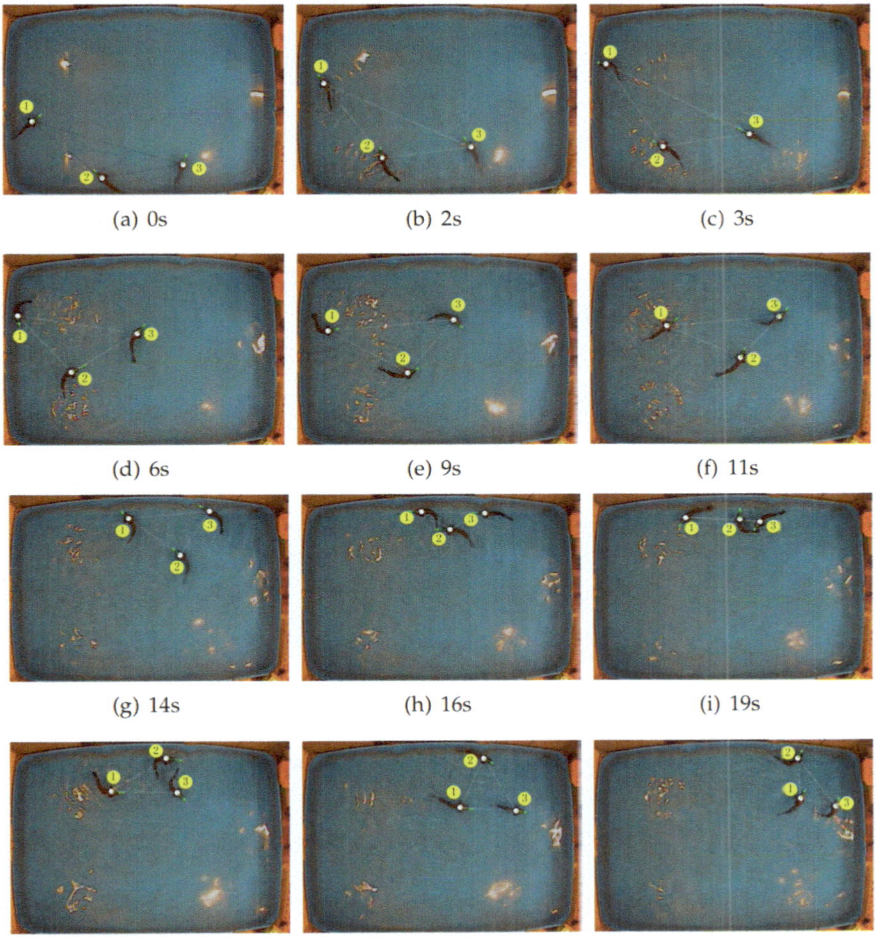

(a) 0s (b) 2s (c) 3s

(d) 6s (e) 9s (f) 11s

(g) 14s (h) 16s (i) 19s

(j) 20s (k) 23s (l) 25s

Fig. 4.9 Flocking experiment scenarios of three robotic fish

the heading angle difference among three robotic fish. Figure 4.10b shows the time response of Θ. From 16 to 20 s, the tremendous change of Θ responds to the boundary collision. Θ asymptotically reaches to a zero value, implying that the heading angles of three robotic fish are consistent. Similarly, the time response of the total potential of three robotic fish is shown in Fig. 4.10c. The total potential presents a decreased trend from an overall point of view, except for some small turnarounds emerging around 4 and 9 s. These turnarounds may be caused by external uncertain disturbances such as water-wave disturbances. Finally, the total potential approximatively converges to a stable value.

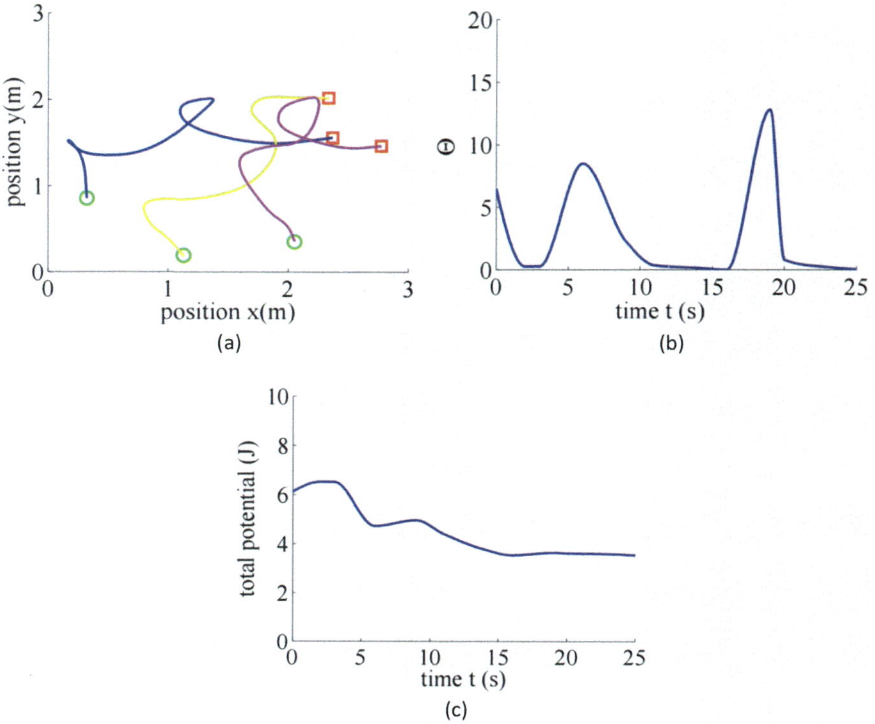

Fig. 4.10 Analysis on experimental results. **a** Trajectories. **b** Energy function of heading angle. **c** Total potential

4.2.4.6 Discussion

The distributed control protocol (4.9) only involves relative position and relative velocity between neighbors. The proposed flocking algorithm would keep working in the presence of differences between the current and the desired relative states among neighbors, no matter these differences are caused by dynamic actuation or uncertain factors. Besides, for multi-agent systems, a distributed algorithm requires lower computation capability and less communication information than a centralized algorithm. These characteristics of the proposed flocking algorithm make it become an effective and practical method to realize dynamic formation control of multiple robotic fish.

However, due to the water-wave disturbances and the bounded experiment field, it is not easy for three robotic fish to maintain a desired formation. Because once they approach to even collide with the boundary of the swimming pool, the formation is commonly destroyed. At the same time, the tracking system of the experiment platform is easy to lose targets, which leads to the interrupt of the experiment process. These difficulties caused by the limitation of external conditions do not deny the feasibility of the proposed control algorithm. Instead, they exactly illustrate the resistant-disturbance capability of the proposed control algorithm to some extent.

Besides, fish must have certain capabilities to participate in the flocking task and maintain the aggregation behavior. For example, it must be able to keep up with the speed of the other fish in the group. Otherwise, it will fall behind and be eaten by predators. Except for motion capabilities, interaction capabilities are also necessary for rapidly making reactions according to the behavior of their neighbors.

4.3 Summary

A distributed control algorithm has been proposed to solve the flocking problem of multiple robotic fish which swims in two-dimensional Euclidean space. By using the nearest-neighbor interaction rules, robotic fish communicate with each other under switching topologies with undirected information flow. The non-holonomic robotic fish with simple structure and weak intelligence is simplified to be an extended second-order unicycle model. Suppose that the initial interaction network is an undirected connected graph. The stability of the closed-loop system is proved by applying LaSalle-Krasovskii invariance principle. Simulation results validate the effectiveness of the flocking algorithm. Finally, experiment results further demonstrate the feasibility of the flocking algorithm for multiple robotic fish, in the presence of field limitations and uncertain disturbances.

References

1. Reynolds, C. W. (1987). Flocks, herds, and schools: A distributed behavioral model. *Computational Graphics, 21*, 25–34.
2. Leonard, N. E., & Fiorelli, E. (2001). Virtual leaders, artificial potentials and coordinated control of groups. In *Proceedings of the 40th IEEE Conference on Decision Control* (Vol. 3, pp. 2968–2973).
3. Olfati-Saber, R. (2006). Flocking for multi-agent dynamic systems: Algorithms and theory. *IEEE Transactions on Automatic Control, 51*, 401–420.
4. Tanner, H. G., Jadbabaie, A., & Pappas, G. J. (2007). Flocking in fixed and switching networks. *IEEE Transactions on Automatic Control, 52*, 863–868.
5. Luo, X., Liu, D., Guan, X., & Li, S. (2012). Flocking in target pursuit for multi-agent systems with partial informed agents. *IET Control Theory and Applications, 6*, 560–569.
6. Jadbabaie, A., Lin, J., & Morse, A. S. (2003). Coordination of groups of mobile autonomous agents using nearest neighbour rules. *IEEE Transactions on Automatic Control, 48*, 988–1001
7. Dimarogonas, D. V., Loizou, S. G., Kyriakopoulos, K. J., & Zavlanos, M. M. (2006). A feedback stabilization and collision avoidance scheme for multiple independent non-point agents. *Automatica, 42*, 229–243.
8. Su, H., Chen, G., Wang, X., & Lin, Z. (2010). Adaptive flocking with a virtual leader of multiple agents governed by nonlinear dynamics. *29th Chinese Control Conference, 2010*, 5827–5832.
9. Sabattini, L., Secchi, C., & Fantuzzi, C. (2011). Arbitrarily shaped formations of mobile robots: Artificial potential fields and coordinate transformation. *Autonomous Robots, 30*, 385–397.

10. Savkin, A. V., & Teimoori, H. (2009). Decentralized formation flocking and stabilization for networks of unicycles. In *Proceedings of the 48th IEEE Conference on Decision Control and the 28th Chinese Control Conference* (pp. 984–989).
11. Gu, D., & Wang, Z. (2009). Leader-follower flocking: Algorithms and experiments. *IEEE Transactions on Control Systems Technology, 17*, 1211–1219.
12. Regmi, A., Sandoval, R., Byrne, R., Tanner, H., & Abdallah, C. T. (2005). Experimental implementation of flocking algorithms in wheeled mobile robots. *Proceedings of the 2005 American Control Conference, 7*, 4917–4922.
13. Su, H., Wang, X., & Chen, G. (2010). Rendezvous of multiple mobile agents with preserved network connectivity. *Systems and Control Letters, 59*, 313–322.
14. Falconi, R., Sabattini, L., Secchi, C., Fantuzzi, C., & Melchiorri, C.: A graph based collision-free distributed formation control strategy. In *18th International Federation of Automatic Control World Congress.*
15. Klein, D. J., Bettale, P. K., Triplett, B. I., & Morgansen, K. A. (2008). Autonomous underwater multivehicle control with limited communication: Theory and experiment. In *International Federation of Automatic Control Workshop on navigation, guidance and control of underwater vehicles.*
16. Weihs, D. (1973). Hydromechanics of fish schooling. *Nature, 241*, 290–291.
17. Low, K. H. (2006). Locomotion and depth control of robotic fish with modular undulating fins. *International Journal of Automation and Computing, 4*, 348–357.
18. Zhang, D., Wang, L., Yu, J., & Tan, M. (2007). Coordinated transport by multiple biomimetic robotic fish in underwater environment. *IEEE Transactions on Control Systems Technology, 15*, 658–671.
19. Zavlanos, M. M., Jadbabaie, A., & Pappas, G. J. (2007). Flocking while preserving network connectivity. In *46th IEEE Conference on Decision Control* (pp. 2919–2924).
20. Yu, J., Wang, L., Shao, J., & Tan, M. (2007). Control and coordination of multiple biomimetic robotic fish. *IEEE Transactions on Control Systems Technology., 15*, 176–183.
21. Shao, J., Wang, L., Yu, J. (2008). Development of an artificial fish-like robot and its application in cooperative transportation. *Control Engineering Practice, 16*, 569–584.
22. Warburton, K., & Lazarus, J. (1991). Tendency distance models of social cohesion in animal groups. *Journal of Theoretical Biology, 150*, 473–488.
23. Ren, W. (2007). Distributed attitude alignment in spacecraft formation flying. *International Journal of Adaptive Control and Signal Processing, 21*, 95–113.
24. Manfredi, S. (2013). Design of a multi-hop dynamic consensus algorithm over wireless sensor networks. *Control Engineering Practice, 21*, 381–394.
25. Khalil, H. K. (2002). *Nonlinear systems* (3rd ed.). Prentice Hall.
26. Sfakiotakis, M., Lane, D. M., & Davies, J. B. C. (1999). Review of fish swimming modes for aquatic locomotion. *IEEE Journal of Oceanic Engineering, 24*, 237–252.
27. Barrett, D., Triantafyllou, M., Yue, D. K. P., Grosenbaugh, M. A., & Wolfgang, M. J. (1999). Drag reduction in fish-like locomotion. *Journal of Fluid Mechanics, 392*, 183–212.
28. Ren, J. (2012). *Collision simulation on robotic fish two-dimensional platform.* Master's thesis, Peking University.

Chapter 5
Applications and Countermeasures for Distributed Coordination Control Systems with Constant Leader Velocity

5.1 Typical Application Scenarios

5.1.1 Maintenance Robots

Imagine a scenario in a lush forest where inspection workers navigate rugged terrains with drones to inspect transmission towers. Due to the considerable distance between towers, much time is spent traversing the jungle. Even utilizing drones, an individual can inspect only 2–3 peaks per day, covering approximately 20 towers.

However, if an inspection worker coordinates with 4 drones, a single peak and its 20 towers can be surveyed in just 40 min, completing the entire inspection in about 3 h.

Clearly, the operational efficiency of multiple drones working in coordination surpasses that of a single drone. Multiple drones performing synchronized tasks mean larger areas can be covered in shorter periods. This requires a scalable and obstacle-avoidance algorithm for effective drone operation, significantly enhancing inspection efficiency. In such cases, swarm strategies are the preferred approach.

5.1.2 Underwater Search and Rescue

In the critical and time-sensitive domain of Underwater Search and Rescue (USAR), distributed cooperative control systems play a pivotal role. In these scenarios, a team of autonomous underwater vehicles (AUVs) or remotely operated vehicles (ROVs) are deployed to perform tasks such as locating wreckage, identifying and rescuing survivors, or delivering essential supplies. The complexities of underwater navigation, combined with the often-uncertain and hazardous environment, demand robust and flexible strategies. These systems leverage advanced algorithms for path planning, obstacle avoidance, and communication, ensuring effective area coverage

Y. Jia, *Distributed Coordination Control System*,
https://doi.org/10.1007/978-981-96-0365-7_5

and data gathering. Implementing a multi-agent approach not only accelerates the search process but also enhances the safety and efficiency of the operations, proving invaluable in mitigating the challenges associated with underwater search and rescue missions.

SwarmDiver is a type of micro-sized Unmanned Surface Vessel (USV) primarily designed for underwater search and rescue missions. Each SwarmDiver measures 750 mm in length and weighs 1.7 kg, equipped with sensors for defense, research, and surveillance purposes. Powered by two brushless DC motors driving propellers, it can reach a maximum speed of 2.2 m/s, with a battery life of 2.5 h. The SwarmDiver has a range of 7 km, specifically engineered for wave environments, capable of measuring temperature and pressure, with GPS tracking accuracy up to 1 m. When operating on the surface, they can communicate with each other via a radio controller.

The system supports single-person operation of the swarm, with algorithms designed for ease of use, supporting various formation types and offering one-click recall. The SwarmDiver can dive to a maximum depth of 50 m, although operators must wait for it to resurface to receive the data collected. This makes it a valuable asset for complex and time-sensitive underwater missions, leveraging the power of collective operation to enhance search and rescue capabilities in marine environments.

5.1.3 Remote Sensing and Earth Observation

Remote Sensing and Earth Observation represent a substantial application domain for distributed cooperative control systems, particularly with the utilization of Unmanned Aerial Vehicles (UAVs) or satellites forming a coordinated network. These systems are instrumental in gathering data across vast and often inaccessible regions, monitoring environmental changes, mapping terrains, and assessing natural disasters.

In such applications, each agent, equipped with high-resolution cameras, sensors, and communication modules, works in a distributed manner to cover large areas systematically. The integration of advanced imaging technologies and real-time data processing allows for the detailed analysis of physical, chemical, and biological conditions of the Earth's surface.

The cooperative nature of these systems enables a comprehensive approach to Earth observation. By working in a swarm, UAVs or satellites can adjust their positions dynamically to optimize coverage and data collection, share tasks and data, and provide redundancy, thereby increasing the reliability and resolution of the observations. These capabilities are crucial for real-time monitoring of environmental phenomena, precision agriculture, urban planning, and disaster response, where timely and detailed information can significantly influence the outcomes.

Drones offer cost, technological, and convenience advantages for nearly all industries with geographic information needs. In scenarios involving a single drone, extensive Earth observation can be time-consuming. By integrating distributed cooperative control technology, especially for time-varying tasks like tracking uncertain ground targets, it's possible to control a large fleet of drones with minimal personnel, addressing the challenges of time and efficiency.

Traditionally, tasks like agricultural mapping and protection required manual labor for marking points on the ground, a process that's labor-intensive and time-consuming. Acquiring farm boundaries manually, step by step, is especially arduous for dispersed, small, or irregularly shaped plots, leading to difficulties in measurement and calculation and often imprecise results. JiXian agricultural remote sensing drones can rapidly generate high-definition 3D digital maps of farmlands, providing precise 3D flight paths for drone-based agricultural tasks in orchards, tea gardens, fields, and rice paddies. This technology resolves the slow pace of marking for protection mapping and the difficulties of manually encircling areas and estimating tree heights and sizes. It ensures the safety of flight operations, reduces mapping costs and agricultural drone operating expenses, and enhances the efficiency of crop protection.

In large-scale, unified prevention and control operations, the collaboration of numerous remote sensing agricultural drones can further improve the efficiency of crop protection tasks. The technological advancements in agricultural remote sensing drones not only reduce reliance on manual mapping, decreasing mapping costs and enhancing operational efficiency, but also rapidly acquire farmland information. This provides scientific, precise decision-making support for agricultural production, marking a significant leap forward in the application of drones in the industry.

5.2 Countermeasure Research

The solutions discussed in this chapter are primarily targeted at multi-agent systems with the following characteristics: the agents within the system are divided into two categories. One category consists of leaders, who are aware of the flight trajectory that the group must follow and the destination's location. These leaders move towards the destination at a constant speed along the desired trajectory. The other category consists of followers, who do not need to know the overall goal. Each follower possesses equal communication capabilities, equal positioning abilities, and a consistent decision-making mechanism.

The primary goal of the solutions proposed in this chapter is to achieve a group coordination task where the multi-agent system moves along a specified trajectory. This means that all agents will move at the same rate as the leaders, in the same direction, and maintain a compact formation around the leaders. Throughout the process, the agents will avoid collisions with each other, and the formation will remain robust. Even if the formation is temporarily disrupted by external disturbances, it will quickly and automatically adjust back to a compact formation and maintain it. Additionally, when it is necessary for the agents to perform coordination tasks in a specified geometric formation, this chapter also provides a formation control solution based on swarm strategies.

This chapter uses robotic fish as the subject of study and provides specific solutions.

5.2.1 Introduction

Collective behavior seems to be a risk-averse mode in animals' abilities to respond to migration, foraging, or predator avoidance [1]. Flocking is a form of collective behavior of large number of interacting agents that move with a common velocity while maintain a cohesive or geometric formation [2, 3]. In recent years, flocking problem of multi-agent system has become a subject of significant interest in the field of control theory, due to its potential application in mobile sensor network, multi-robot systems, etc. [4–9].

Migrating flocking, known as leader–follower flocking, is a special flocking phenomenon that can be widely observed among schools of fish, flocks of birds, or other social animals in nature. In the colony, only one or a few individuals could know the global information, such as the location of a food source or the route of a migration, while the others adjust their behaviors according to the local information from their neighbors.

The leader–follower mechanism has been widely employed in formation control research for multi-robot systems [10–19]. For instance, Yang et al. utilized the back-stepping method to coordinate the movement of followers, ensuring that their positions and orientations relative to the leader reach the desired state [15]. Mariottini et al. addressed the vision-based localization problem in leader–follower formation control for unicycles equipped with panoramic cameras [16]. Ailon et al. proposed a leader–follower control strategy based on smooth properties and virtual agents, enabling a group of mobile robots to maintain a desired formation along a time-parameterized path [19]. Besides these control methods, with the rapid development of sensing and communication technologies, mature flocking algorithms for multi-agent systems now possess the capability to achieve coordinated control in multi-robot systems.

The existence of a virtual leader is a common assumption in the current literature on flocking control [6, 20–26]. Unlike these virtual leaders, the leaders in nature or engineering applications are often real individuals within the group, responsible for participating in energy distribution and formation construction within the system, thereby introducing new challenges to the study of flocking control [15, 17, 19].

For example, a multi-leader following strategy was presented to deal with the flocking problem of multiple agents with double integrator dynamics [27], which cannot be directly applied to modern vehicles with complex physical constraints. Suppose that interconnection between leader and follower was bidirectional, a leader–follower flocking problem of wheeled mobile robots was studied, during which the position of flocking center to keep the group connected was estimated by consensus algorithm [28]. Besides, artificial potential field method is another common method to solve the connectivity preservation problem [29]. Furthermore, the concrete results regarding the connection of formation control and flocking behavior of multiple non-holonomic kinematic agents have been given, but no leaders were involved [5, 30].

The purpose of this paper is to deal with the leader–follower cohesive flocking problem and the leader–follower formation flocking problem for multiple robotic fish under special consideration of their kinematic constraints, their communication capability, and their control mechanism. Each robotic fish swims in the water surface. There exists only one leader. The leader acts independently of the followers. Thus, the interconnection between leader and follower is unidirectional [31]. In addition, since a stable formation of the group can be maintained only when the external control input of the leader is zero, we just consider the case that the leader swims with a constant forward speed and a constant rotational speed.

The flocking behavior of the robotic fish has three essential features, that is, velocity matching, collision avoidance, and flock centering [5]. Based on the combination of consensus protocol and potential function, a distributed cohesive flocking algorithm is designed for the robotic fish system consisting of one leader and several followers to structure a cohesive formation. The stability of the closed-loop system is analyzed by means of graph theory and LaSalle-Krasovskii invariance principle. In order to extend its application areas, the proposed cohesive flocking algorithm is further revised to achieve the formation flocking task. Two flocking experiments are performed by three robotic fish swimming in the water surface. Finally, several simulation examples are conducted to verify the effectiveness of the proposed algorithms for a ten-agent system.

5.2.2 *Model and Problem Statement*

To simplify the problem, starting from this chapter, the robotic fish model will not include external disturbance terms. The kinematic characteristics of robotic fish $i (i = 1, \ldots, N)$ can be described by the following model:

$$
\begin{aligned}
\dot{x}_i(t) &= v_i(t)\cos\theta_i(t) - \omega_i(t) l_i \sin\theta_i(t) \\
\dot{y}_i(t) &= v_i(t)\sin\theta_i(t) + \omega_i(t) l_i \cos\theta_i(t) \\
\dot{\theta}_i(t) &= \omega_i(t) \\
\dot{v}_i(t) &= a_i(t) \\
\dot{\omega}_i(t) &= b_i(t)/l_i
\end{aligned}
\tag{5.1}
$$

where $p_i(t) = \left[x_i(t), y_i(t)\right]^T \in \mathbb{R}^2$ is the position vector of the robotic fish i at time t, $\theta_i(t) \in [0, 2\pi)$ is the heading angle measured from the x-axis in the anticlockwise rotation, $q_i(t) = [v_i(t), l_i\omega_i(t)]^T = [v_i(t), \vartheta_i(t)]^T$ is the velocity vector, $v_i(t) \in \mathbb{R}$

Fig. 5.1 Schematic model for the robotic fish i

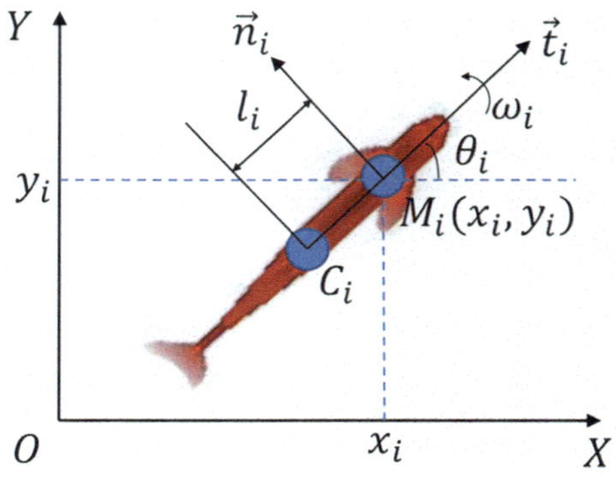

is the forward speed, $\omega_i(t) \in \mathbb{R}$ is the rotational speed, $l_i \in \mathbb{R}^+$ is the distance between the geometrical center C_i and the mass center M_i, $\vartheta_i(t) = l_i\omega_i(t) \in \mathbb{R}$ is the tangential speed, $a_i(t) \in R$ is the translational acceleration, and $b_i(t) \in R$ is the rotational acceleration. We do not consider the individual differences, and thus assume that $l_i = l_d$, for $i = 1, \dots, N$, where l_d is a positive constant. Note that in this chapter, unless otherwise specified, the velocity of the robotic fish is bounded and all variables are time-varying.

Figure 5.1 gives the graphical representation of the fish model. Therein, $\vec{t}_i = [\cos\theta_i, \sin\theta_i]^T$ and $\vec{n}_i = [-\sin\theta_i, \cos\theta_i]^T$ are two unit vectors orthogonal to each other. With $p_i(t) = \left[x_i(t), y_i(t)\right]^T$ and $q_i(t) = [v_i(t), l_i\omega_i(t)]^T$, some equations in (5.1) can be merged as the following matrix from

$$\begin{aligned} \dot{p}_i &= H_i^T q_i \\ \dot{q}_i &= u_i \end{aligned} \tag{5.2}$$

where $H_i = \left[\vec{t}_i, \vec{n}_i\right]^T = \begin{bmatrix} \cos\theta_i & \sin\theta_i \\ -\sin\theta_i & \cos\theta_i \end{bmatrix}$, and $u_i = [a_i, b_i]^T$. Besides, u_i is taken as the control input of robotic fish i.

The robotic fish system under consideration consists of N robotic fish. There are only one leader and N − 1 followers. Let leader set be $\mathbb{L} = \{1\}$ and follower set be $\mathbb{F} = \{2, \dots, N\}$. If an agent has external control input, we call it a leader; otherwise, we call it a follower. We suppose that for a follower, interconnection with the leader is unidirectional, while interconnection with the follower is bidirectional [32]. Let N_i(t) denote the neighbor set of the follower $i \in \mathbb{F}$ at time t, and the initial neighbor set of the follower i is defined as

$$N_i(0) = \left\{j | p_i(0) - p_j(0) < D, j = 1, \dots, N, j \neq i\right\} \tag{5.3}$$

where D > 0 is a constant, and $\|\cdot\|$ is the Euclidean norm.

The interaction network G(t) is a dynamic directed graph consisting of a vertex set $\nu = \{1, \dots, N\}$ indexed by robotic fish and a time-varying edge set $\varepsilon(t) =$

$\{(i,j)|(i,j) \in \mathbb{F} \times \nu, j \in N_i(t)\}$. Therein, the followers' interaction network $\hat{G}(t)$ with vertex set F and edge set $\hat{\varepsilon}(t) = \{(i,j)|(i,j) \in \mathbb{F} \times \mathbb{F}, j \in N_i(t)\}$ is an undirected graph. In order to clarify the neighbor relationship, we introduce the adjacency matrix $A_N(t)$ of the graph G(t) and the Laplacian matrix $L_{N-1}(t)$ of the graph $\hat{G}(t)$. $A_N(t) = \left[\omega_{ij}(t)\right]_{N\times N}$ is defined as

$$\omega_{ij}(t) = \begin{cases} 1, i = 1, j = 1 \\ 1, i \in F, j \in N_i(t) \\ 0, \textit{otherwise} \end{cases} \tag{5.4}$$

and $L_{N-1}(t) = \left[L_{ij}(t)\right]_{(N-1)\times(N-1)}$ is given by

$$l_{ij}(t) = \begin{cases} -\omega_{ij}, i \neq j \\ \sum\limits_{k=2,k\neq i}^{N} \omega_{ik}, i = j \end{cases} \tag{5.5}$$

As $\hat{G}(t)$ is an undirected graph, the Laplacian matrix $L_{N-1}(t)$ is symmetric and positive semi-definite.

Provided that the external input of the leader is zero, the flocking problem involved in this paper is how to design control protocols to make the followers track the leader's speeds and to approach the equilibrium distances between any two neighbors, during which collisions are avoided as much as possible.

The size of the real robotic fish is about 0.45 m $\times$ 0.05 m $\times$ 0.12 m (length $\times$ width $\times$ height), and its minimum turning radius is $R_{min} = 20cm$. Suppose that there is no distance between any two robotic fish less than $2R_{min}$ at the initial time, a loose condition is considered to guarantee the collision avoidance, that is, any two robotic fish could avoid collision by swimming with the minimum turning radius as $\|p_{ij}\| \to 2R_{min}^{+}$.

Given that G(0) is a leader–follower connected graph, which means that there exists at least one directed path from the leader to each follower. In order to preserve the connectivity of the interaction network G(t), the hysteresis adding new edges to the network is introduced [29][32], such that.

(1) if $(i,j) \in \varepsilon(t^-)$ and $\|\hat{p}_i(t) - \hat{p}_j(t)\| < 2D$, where $\hat{p}_i(t) = p_i(t) - p_i^*(t)$ and $p_i^*(t) = \int_0^t \omega_{i1}(t) H_i^T(t) q_1 d\tau$ (i = 1, ..., N), then (i, j) $\in$ ε(t), for t > 0;
(2) if $(i,j) \notin \varepsilon(t^-)$ and $\|p_i(t) - p_j(t)\| < D$, then (i, j) $\in$ ε(t), for t > 0.

Define that the interaction network G(t) switches at t_p, $p = 1, 2, \ldots$. Thus, G(t) is a fixed graph in each nonempty, bounded, and contiguous time-interval $[t_{r-1}, t_r)$, where $r = 0, 1, \ldots$.

5.2.3 Distributed Coordination Control Algorithm Design and Stability Analysis

5.2.3.1 Cohesive Flocking

1 Algorithm Design

Without considering time delay, the leader's control protocol is essentially its external control input, which is given by:

$$u_1(t) = \begin{bmatrix} a_1(t) \\ b_1(t) \end{bmatrix} = 0 \tag{5.6}$$

provided that the leader does a curve motion with a constant forward speed v_1 and a constant rotational speed $\omega_1 (\omega_1 \neq 0)$ Note that $q_1 = [v_i, l_d \omega_1]$ is a constant vector. Thus, the control protocol of the follower i(i $= 2, \ldots,$ N) consists of two components:

$$u_i(t) = \begin{bmatrix} a_i(t) \\ b_i(t) \end{bmatrix} = \alpha_i(t) + \beta_i(t) \tag{5.7}$$

where $\alpha_i(t)$ is the consensus term, and $\beta_i(t)$ is the gradient-based term. The first part $\alpha_i(t)$ aims at making all the robotic fish move with consistent velocities by consensus algorithms. The second part $\beta_i(t)$ mainly affects the relative positions between individuals in fish schooling through the negated gradients of the attractive/repulsive functions [20].

Let $\hat{v}_i = v_i - \omega_{i1} v_1$, $\hat{\theta}_i = \theta_i - \omega_{i1}\theta_1$, $\hat{\omega}_i = \omega_i - \omega_{i1}\omega_1$, and $\hat{q}_i = q_i - \omega_{i1} q_1$, $i = 1, \ldots, N$. Coupled with (5.2), we have

$$\dot{\hat{p}}_i = \dot{p}_i - \dot{p}_i^* = H_i^T q_i - \omega_{i1} H_i^T q_1 = H_i^T \hat{q}_i, i = 1, \ldots, N \tag{5.8}$$

Then, the control protocol for follower i(i $= 2, \ldots$ N) is designed by

$$\begin{aligned} a_i(t) &= -\sum_{j \in N_i(t)} (\hat{v}_i - \hat{v}_j) - \sum_{j \in N_i(t)} \left(\dot{\hat{p}}_i^T - \dot{\hat{p}}_j^T\right)\vec{t}_i - \sum_{j \in N_i(t)} \nabla_{\hat{p}_{ij}} V\left(\|\hat{p}_{ij}\|\right)^T \vec{t}_i \\ b_i(t) &= -l_d \sum_{j \in N_i(t)} \left(\hat{\theta}_i - \hat{\theta}_j\right) - l_d \sum_{j \in N_i(t)} \left(\hat{\omega}_i - \hat{\omega}_j\right) \\ &\quad - \sum_{j \in N_i(t)} \left(\dot{\hat{p}}_i^T - \dot{\hat{p}}_J^T\right)\vec{n}_i - \sum_{j \in N_i(t)} \nabla_{\hat{p}_{ij}} V\left(\|\hat{p}_{ij}\|\right)\vec{n}_i \end{aligned} \tag{5.9}$$

where $\hat{p}_{ij} = \hat{p}_i - \hat{p}_j$, and $\nabla_{\hat{p}_{ij}} V\left(\hat{p}_{ij}\right)$ is the gradient of an artificial potential function $V\left(\hat{p}_{ij}\right)$ with the following definition:

Definition 5.1 (Potential function) Potential $V(\hat{p}_{ij})$ is a differentiable, nonnegative, radially unbounded function of the Euclidean norm $\hat{p}_{ij}$ between agents i and j, such that

(1) $V(\hat{p}_{ij}) \to \infty$ as $\hat{p}_{ij} \to 0$;
(2) $V(\hat{p}_{ij}) \to \infty$ as $\hat{p}_{ij} \to 2D$;
(3) $V(\hat{p}_{ij})$ attains its unique minimum when the Euclidean norm $\hat{p}_{ij}$ equals to a certain value between 0 and 2D.

Besides, the total potential of the system is

$$V = \sum_{i \in \mathbb{F}} \sum_{j \in N_i} V(\hat{p}_{ij}) + \sum_{j \in N_l} V(\hat{p}_{1j}) \tag{5.10}$$

Assume that $q_i^* = [0, l_d\theta_i]^T$ and $\hat{q}_i^* = q_i^* - \omega_{i1} q_1^* = \left[0, l_d\hat{\theta}_i\right]^T$, $i = 1, \ldots, N$. From (5.2), one gets $\dot{q}_i = u_i = [a_i, b_i]^T$. Then, the control protocol (5.9) can be rewritten as

$$\begin{aligned} \dot{q}_i = &- \sum_{j \in N_i(t)} \left(\hat{q}_i^* - \hat{q}_j^*\right) - \sum_{j \in N_i(t)} (\hat{q}_i - \hat{q}_j) \\ &- H_i \sum_{j \in N_i(t)} \left(\dot{\hat{p}}_i - \dot{\hat{p}}_j\right) - H_i \sum_{j \in N_i(t)} \nabla_{\hat{p}_{ij}} V(\hat{p}_{ij}) \end{aligned} \tag{5.11}$$

(2) **Constraint Conditions**

As shown in Fig. 5.2, the specific potential function is given by

$$V(\hat{p}_{ij}) = \frac{b}{\hat{p}_{ij}^2} - a ln\left(4D^2 - \hat{p}_{ij}^2\right) + c \tag{5.12}$$

where a and b are positive constants, and c is a constant. Therein, the first part $\frac{b}{\hat{p}_{ij}^2}$ is the repulsive potential, the second part $-a ln\left(4D^2 - \hat{p}_{ij}^2\right)$ is the attractive potential, and the third constant item c is just used to guarantee the potential function positive at all times.

Then, the gradient of the potential function is obtained by

$$\nabla_{\hat{p}_{ij}} V(\hat{p}_{ij}) = 2\hat{p}_{ij} \left(\frac{b}{\hat{p}_{ij}^2} - \frac{a}{4D^2 - \hat{p}_{ij}^2} \right) \tag{5.13}$$

where $\frac{b}{\hat{p}_{ij}^2}$ represents the magnitude of the repulsion term with a short range, and $\frac{a}{4D^2 - \hat{p}_{ij}^2}$ represents the magnitude of the attraction term with a long range. The attraction term and repulsion term act in opposite directions. There is a unique distance $\hat{p}_{ij} =$

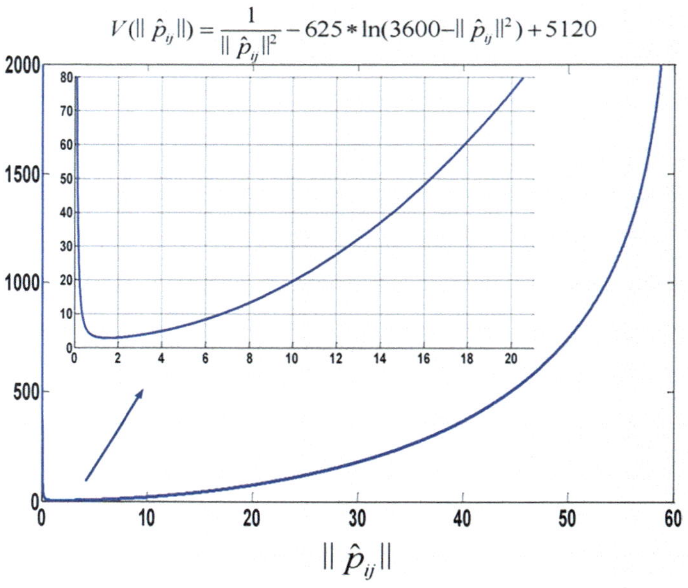

Fig. 5.2 The potential function (a = 625, b = 1, c = 5120, D = 30)

$\sqrt{\frac{\sqrt{b^2+16abD^2}-b}{2a}}$, at which the attraction and the repulsion balance [33], i.e., $\frac{b}{\hat{p}_{ij}^2} = \frac{a}{4D^2-\hat{p}_{ij}^2}$. When $a = 625$, $b = 1$, and $D = 30$, we have $\sqrt{\frac{\sqrt{b^2+16abD^2}-b}{2a}} = 1.5489$.

Specially, it should be noted that parameters a and b at least satisfy the constraint condition

$$4R_{min}^2 < \frac{\sqrt{b^2 + 16abD^2} - b}{2a} < D^2 \tag{5.14}$$

Besides, in order to ensure the collision avoidance between any two robotic fish, the repulsion force is required to be large enough that makes the robotic fish under the control protocol (5.9) swim with minimum turning radius as $\|\hat{p}_{ij}\| = 2R_{min} + \left\|p_{ij\,max}^*\right\|$. Here, $\left\|p_{ij\,max}^*\right\|$ is the upper bound of $\left\|p_{ij}^*\right\|$.

(3) **Stability Analysis**

In this subsection, we prove that the robotic fish system (5.1) applying the control protocol (5.9) will be asymptotically stable, under some appropriate constraints on the potential function and the velocity of the leader. To this end, LaSalle-Krasovskii invariance principle and graph theory are used. So far, we are in a position to state our main result on the leader–follower cohesive flocking problem.

Theorem 5.1 (Leader–follower cohesive flocking) *Consider a system of N agents with dynamics (5.1). The leader and followers are respectively steered by control protocols (5.6) and (5.9). Suppose that the initial interaction topology of the system is a leader–follower connected graph. Then the following statements hold* (Fig. 5.3):

1. *No collision happens between neighbors.*
2. *The connectivity of the interaction topology is preserved at all times.*
3. *The forward speed, the rotational speed, and the heading angle of each follower asymptotically become the same as those of the leader.*
4. *The system approaches a cohesive configuration that minimizes the total potential V.*

Proof Consider the system (5.1) with control input (5.6) and (5.9) on time interval $[t_r, t_{r+1})$, where the interaction network of the robotic fish system is fixed. Due to $\hat{q}_i = q_i - w_{i1}q_1$, where w_{i1} is constant and $q_1 = [\upsilon_1, l_d\omega_1]^T$ is a constant vector, we have that $\dot{\hat{q}}_i = \dot{q}_1$. According to (5.11), it is easy to obtain that

$$\dot{\hat{q}}_i = -\sum_{j\in N_i}\left(\hat{q}_i^* - \hat{q}_j^*\right) - \sum_{j\in N_i}(\hat{q}_i - \hat{q}_j) - H_i\sum_{j\in N_i}\left(\dot{\hat{p}}_i - \dot{\hat{p}}_i\right) - H_i\sum_{j\in N_i}\nabla_{\hat{p}_{ij}}V\left(\hat{p}_{ij}\right) \tag{5.15}$$

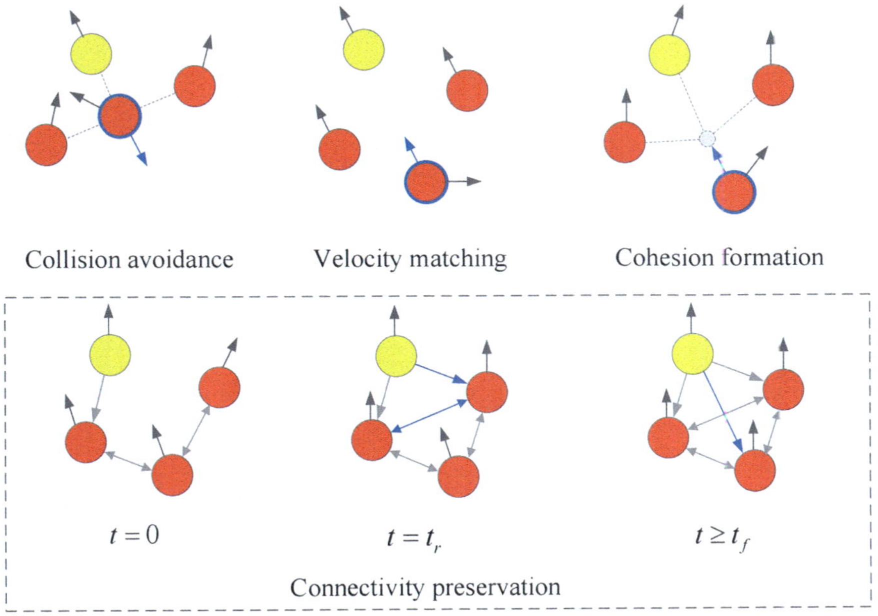

Fig. 5.3 The conclusions of Theorem 5.1

Let $\tilde{p} = \left[\hat{p}_{11}^T, \ldots, \hat{p}_{1N}^T, \ldots, \hat{p}_{N1}^T, \ldots, \hat{p}_{NN}^T\right]^T$, $\hat{q} = \left[\hat{q}_2^T, \ldots, \hat{q}_N^T\right]^T$, and $\hat{\theta} = \left[\hat{\theta}_2, \ldots, \hat{\theta}_N\right]^T$. The set of followers who have one leader neighbor is expressed by

$$N_l = \{i | w_{i1} = 1, i \in \mathbb{F}\} \tag{5.16}$$

Consider the following energy function as the common Lyapunov function

$$E\left(\tilde{p}, \hat{q}, \hat{\theta}\right) = \frac{1}{2}V + \frac{1}{2}\sum_{i\in\mathbb{F}} \hat{q}_i^T \hat{q}_i + \frac{1}{2}\Theta \tag{5.17}$$

where $\Theta = l_d^2 \sum_{i\in\mathbb{F}} \sum_{j\in N_i} \hat{\theta}_i\left(\hat{\theta}_i - \hat{\theta}_j\right)$. With $\hat{\theta}_1 = 0$, we have $\dot{\Theta} = 2l_d^2 \sum_{i\in\mathbb{F}} \sum_{j\in N_i} \dot{\hat{\theta}}_i\left(\hat{\theta}_i - \hat{\theta}_j\right)$. Then the derivative of $E\left(\tilde{p}, \hat{q}, \hat{\theta}\right)$ w.r.t. the time $t \in \left[t_r, t_{r+1}\right]$ is

$$\begin{aligned}
\frac{dE}{dt} &= \frac{1}{2}\sum_{i\in\mathbb{F}}\sum_{j\in N_i} \dot{\hat{p}}_{ij}^T \nabla_{\hat{p}_{ij}} V\left(\left\|\hat{p}_{ij}\right\|\right) + \frac{1}{2}\sum_{j\in N_l} \dot{\hat{p}}_{1j}^T \nabla_{\hat{p}_{1j}} V\left(\left\|\hat{p}_{1j}\right\|\right) \\
&\quad + \sum_{i\in\mathbb{F}} \hat{q}_i^T \dot{\hat{q}}_i + l_d^2 \sum_{i\in\mathbb{F}}\sum_{j\in N_i} \dot{\hat{\theta}}_i\left(\hat{\theta}_i - \hat{\theta}_j\right) \\
&= \sum_{i\in\mathbb{F}} \dot{\hat{p}}_i^T \sum_{j\in N_i/\{1\}} \nabla_{\hat{p}_{ij}} V\left(\left\|\hat{p}_{ij}\right\|\right) + \frac{1}{2}\sum_{j\in N_l} \dot{\hat{p}}_{i1}^T \nabla_{\hat{p}_{i1}} V\left(\left\|\hat{p}_{i1}\right\|\right) \\
&\quad + \frac{1}{2}\sum_{j\in N_l} \dot{\hat{p}}_{1j}^T \nabla_{\hat{p}_{1j}} V\left(\left\|\hat{p}_{1j}\right\|\right) + \sum_{i\in\mathbb{F}} \hat{q}_i^T \dot{\hat{q}}_i + l_d^2 \sum_{i\in\mathbb{F}}\sum_{j\in N_i} \dot{\hat{\theta}}_i\left(\hat{\theta}_i - \hat{\theta}_j\right)
\end{aligned} \tag{5.18}$$

Due to $\dot{p}_{ij}^T = -\dot{p}_{ij}^T$ and the symmetric nature of $V\left(\hat{p}_{ij}\right)$, one gets

$$\begin{aligned}
\frac{1}{2}\sum_{j\in N_l} \dot{\hat{p}}_{1j}^T \nabla_{\hat{p}_{1j}} V\left(\left\|\hat{p}_{1j}\right\|\right) &= \frac{1}{2}\sum_{j\in N_l} \dot{\hat{p}}_{j1}^T \nabla_{\hat{p}_{j1}} V\left(\left\|\hat{p}_{j1}\right\|\right) \\
&= \frac{1}{2}\sum_{j\in N_l} \dot{\hat{p}}_{i1}^T \nabla_{\hat{p}_{i1}} V\left(\left\|\hat{p}_{i1}\right\|\right)
\end{aligned} \tag{5.19}$$

Thus,

$$\begin{aligned}
\frac{dE}{dt} &= \sum_{i\in\mathbb{F}} \dot{\hat{p}}_i^T \sum_{j\in N_i/\{1\}} \nabla_{\hat{p}_{ij}} V\left(\left\|\hat{p}_{ij}\right\|\right) + \sum_{j\in N_l} \dot{\hat{p}}_{i1}^T \nabla_{\hat{p}_{i_1}} V\left(\left\|\hat{p}_{i1}\right\|\right) \\
&\quad + \sum_{i\in\mathbb{F}} \hat{q}_i^T \dot{\hat{q}}_i + l_d^2 \sum_{i\in\mathbb{F}}\sum_{j\in N_i} \dot{\hat{\theta}}_i\left(\hat{\theta}_i - \hat{\theta}_j\right)
\end{aligned} \tag{5.20}$$

We have defined that $w_{11} = 1$. According to (5.8), one gets $\dot{\hat{p}}_1 = H_1^T \hat{\mathrm{q}}_1 = H_1^T (q_1 - w_{11} q_1) = 0$. With $\dot{\hat{p}}_{i1}^T = \dot{\hat{p}}_i^T - \dot{\hat{p}}_1^T = \dot{\hat{p}}_i^T$, we have

$$\begin{aligned}\frac{dE}{dt} &= \sum_{i\in\mathbb{F}} \dot{\hat{p}}_i^T \sum_{j\in N_i/\{1\}} \nabla_{\hat{p}_{ij}} V(\|\hat{p}_{ij}\|) + \sum_{i\in N_l} \dot{\hat{p}}_i^T \nabla_{\hat{p}_{i1}} V(\|\hat{p}_{i1}\|) \\ &\quad + \sum_{i\in\mathbb{F}} \hat{q}_i^T \dot{\hat{q}}_l + l_d^2 \sum_{i\in\mathbb{F}} \sum_{j\in N_i} \dot{\hat{\theta}}_l (\hat{\theta}_i - \hat{\theta}_j) \\ &= \sum_{i\in\mathbb{F}} \dot{\hat{p}}_i^T \sum_{j\in N_i} \nabla_{\hat{p}_{ij}} V(\|\hat{p}_{ij}\|) + \sum_{i\in\mathbb{F}} \hat{q}_i^T \dot{\hat{q}}_l + l_d^2 \sum_{i\in\mathbb{F}} \sum_{j\in N_i} \dot{\hat{\theta}}_l (\hat{\theta}_i - \hat{\theta}_j)\end{aligned} \tag{5.21}$$

According to (5.8) and (5.15), dE/dt can be further expressed by

$$\begin{aligned}\frac{dE}{dt} &= \sum_{i\in\mathbb{F}} \hat{q}_i^T H_i \sum_{j\in N_i} \nabla_{\hat{p}_{ij}} V(\hat{p}_{ij}) + l_d^2 \sum_{i\in\mathbb{F}} \sum_{j\in N_i} \dot{\hat{\theta}}_i (\hat{\theta}_i - \hat{\theta}_j) \\ &\quad + \sum_{i\in\mathbb{F}} \hat{q}_i^T \Big(- \sum_{j\in N_i} \left(\hat{q}_i^* - \hat{q}_j^*\right) - \sum_{j\in N_i} (\hat{q}_i - \hat{q}_j) \\ &\quad - H_i \sum_{j\in N_i} \left(\dot{\hat{p}}_i - \dot{\hat{p}}_i\right) - H_i \sum_{j\in N_i} \nabla_{\hat{p}_{ij}} V(\hat{p}_{ij})\Big) \\ &= l_d^2 \sum_{i\in\mathbb{F}} \sum_{j\in N_i} \dot{\hat{\theta}}_i (\hat{\theta}_i - \hat{\theta}_j) - \sum_{i\in\mathbb{F}} \hat{q}_i^T \sum_{j\in N_i} \left(\hat{q}_i^* - \hat{q}_j^*\right) \\ &\quad - \sum_{i\in\mathbb{F}} \hat{q}_i^T \sum_{j\in N_i} (\hat{q}_i - \hat{q}_j) - \sum_{i\in\mathbb{F}} \hat{q}_i^T H_i \sum_{j\in N_i} (\dot{\hat{p}}_i - \dot{\hat{p}}_j) \\ &= l_d^2 \sum_{i\in\mathbb{F}} \sum_{j\in N_i} \dot{\hat{\theta}}_i (\hat{\theta}_i - \hat{\theta}_j) - \sum_{i\in\mathbb{F}} \sum_{j\in N_i} \hat{q}_i^T \left(\hat{q}_i^* - \hat{q}_j^*\right) \\ &\quad - \sum_{i\in\mathbb{F}} \hat{q}_i^T \sum_{j\in N_i} (\hat{q}_i - \hat{q}_j) - \sum_{i\in\mathbb{F}} \dot{\hat{p}}_i^T \sum_{j\in N_i} \left(\dot{\hat{p}}_i - \dot{\hat{p}}_j\right)\end{aligned} \tag{5.22}$$

From $\hat{q}_i^T \left(\hat{q}_i^* - \hat{q}_j^*\right) = l_d^2 \hat{\omega}_i (\hat{\theta}_i - \hat{\theta}_j)\ \dot{\hat{\theta}}_i (\hat{\theta}_i - \hat{\theta}_j) = \hat{\omega}_i (\hat{\theta}_i - \hat{\theta}_j)$, it is easy to obtain that $l_d^2 \sum_{i\in\mathbb{F}} \sum_{j\in N_i} \dot{\hat{\theta}_i} (\hat{\theta}_i - \hat{\theta}_j) = \sum_{i\in\mathbb{F}} \sum_{j\in N_i} \hat{q}_i^T \left(\hat{q}_i^* - \hat{q}_j^*\right) - \sum_{i\in\mathbb{F}} \hat{q}_i^T$. $\frac{dE}{dt}$ can be simplified by

$$\frac{dE}{dt} = - \sum_{i\in\mathbb{F}} \hat{q}_i^T \sum_{j\in N_i} (\hat{q}_i - \hat{q}_j) - \sum_{i\in\mathbb{F}} \dot{\hat{p}}_i^T \sum_{j\in N_i} (\dot{\hat{p}}_I - \dot{\hat{p}}_J) \tag{5.23}$$

Owing to $\hat{q}_1 = 0$, one gets

$$- \sum_{i\in\mathbb{F}} \hat{q}_i^T \sum_{j\in N_i} (\hat{q}_i - \hat{q}_j) = -\hat{\text{q}}^T (L_{N-1} \otimes I_2) \hat{q} - \sum_{i\in N_l} \hat{q}_i^T \hat{q}_i \le 0 \tag{5.24}$$

where I_2 denotes the 2×2 identity matrix, and L_{N-1} is symmetric and positive semi-definite. Similarly, due to $\dot{\hat{p}}_1 = 0$, we have

$$-\sum_{i\in\mathbb{F}} \dot{\hat{p}}_i^T \sum_{j\in N_i} (\hat{p}_i - \hat{p}_j) = -\hat{p}(L_{N-1} \otimes I_2)\dot{\hat{p}}^T - \sum_{i\in\mathbb{F}} \dot{\hat{p}}_i^T \dot{\hat{p}}_i \le 0 \tag{5.25}$$

where $\hat{p} = \left[\hat{p}_2^T, \ldots, \hat{p}_N^T\right]^T$. Hence,

$$\begin{aligned}\frac{dE}{dt} &= -\hat{\mathrm{q}}^T (L_{N-1} \otimes I_2)\hat{q} - \sum_{i\in N_l} \hat{q}_i^T \hat{q}_i \\ &- \dot{\hat{p}}^T (L_{N-1} \otimes I_2)\dot{\hat{p}} - \sum_{i\in\mathbb{F}} \dot{\hat{p}}_i^T \dot{\hat{p}}_i \le 0\end{aligned} \tag{5.26}$$

The initial potential energy of the system is finite, and the initial speeds and initial heading angles of all agents are also finite. Thus, the initial energy $E(\tilde{p}(0), \hat{q}(0), \hat{\theta}(0))$ of the system is finite. Since $dE/dt \le 0$ for each time-interval $[t_r, t_{r+1}), r = 0, 1, \ldots,$ the supremum of E(t) is obviously its initial value $E(\tilde{p}(0), \hat{q}(0), \hat{\theta}(0))$, for all t. Then, the potential energy V(t) of the system is finite. Furthermore, the potential energy $V(\|\hat{p}_{ij}(t)\|)$ between agent i and j is also finite.

(1) Collision Avoidance

According to rule (1) of definition 5.1, one gets $V(\|\hat{p}_{ij}\|) \to \infty$ as $\hat{p}_{ij} \to 0$. However, $V(\|\hat{p}_{ij}\|) \to \infty$ violates the conclusion that $V(\|\hat{p}_{ij}(t)\|)$ remains finite. Thus, $\hat{p}_{ij} > 0$. Consider that $\left\|\left(w_{i1}H_i^T - w_{j1}H_j^T\right)q_1\right\| = \|q_1\|$, we have

$$\begin{aligned}\left\|p_{ij}^*\right\| &= \left\|\int_{t_r}^{t_{r+1}} \left(w_{i1}(t)H_i^T(t) - w_{j1}(t)H_j^T(t)\right)q_1 d\tau\right\| \\ &\le \left\|\left(w_{i1}H_i^T - w_{j1}H_j^T\right)q_1\right\|(t_{r+1} - t_r) \\ &= \|q_1\|(t_{r+1} - t_r)\end{aligned} \tag{5.27}$$

Hence, $\left\|p_{ij}^*\right\|$ is bounded on the bounded time-interval $[t_r, t_{r+1})$, and its upper bound $\left\|p_{ij}^*\right\|_{max}$ is decided by $\|q_1\|$. Provided that the robotic fish under the control strategy (5.9) swims with its minimum turning radius as $\hat{p}_{ij} = 2R_{min} + \left\|p_{ij}^*\right\|_{max}$, which guarantees that robotic fish can avoid collision with each other. Conclusion 1 of Theorem 5.1 is proved.

(2) **Connectivity Preservation**

If for some $(i, j) \in \varepsilon$, $\|\hat{p}_{ij}\| \to 2D$, then according to rule (2) of definition 5.1, one gets $V(\|\hat{p}_{ij}\|) \to \infty$, which violates the conclusion that $V(\|\hat{p}_{ij}\|)$ remains finite. It follows that $\|\hat{p}_{ij}\| < 2D$ for all $(i, j) \in \varepsilon$ and $t \in [t_r, t_{r-1})$. Hence, whenever there is a link between two agents, the link is never lost during each time-interval $[t_r, t_{r+1})$, that is, $\varepsilon(t_r) \subseteq \varepsilon(t_{r+1})$. Note that when $\|\hat{p}_{ij}\|(t) < 2D$, one gets $\|p_{ij}\| < R = 2D + \|p^*_{ij}\|$. It is obvious that $R > D$, which ensures that if an edge $(i, j) \notin \varepsilon$ is added to ε, the associated potential $V(\|\hat{p}_{ij}\|)$ is bounded and hence, so is the new potential V. Consequently, the connectivity of the communication network can be preserved at all times. Specially, since the velocity of the robotic fish is bounded, each robotic fish under the control strategy (5.9) is required to swim with its minimum turning radius as $|p_{ij}| \to (D - 2R_{min})^-$. Conclusion 2 of Theorem 5.1 is proved.

(3) **Velocity Matching**

Assume that there are $m_r \in N$ new edges being added to the interaction network $G(t)$ at switching time t_r(r $= 1, 2, \ldots$). We have supposed that the initial interaction topology of the system is a leader–follower connected graph. Let $G(0)$ denote the initial communication topology of the system, and Gc denote the set of all leader–follower connected graphs on the vertices v. The proposed control protocol can guarantee that the sequence of switching topologies G_{r+1} within $[t_r, t_{r+1})$ consists of such graphs satisfying $G_{r+1} \in G_c$. The number of the vertices is finite, thus G_c is a finite set. Assume that there are at most $M \in N$ new edges that can be added to the initial communication topology $G(0)$. Clearly, $0 < m_r \leq M$ and $r \leq M$. Therefore, the number of switching times of the system is finite, and the interaction network $G(t)$ eventually becomes fixed. Suppose that the last switching time is t_f, the following discussions are restricted on the time interval $[t_f, \infty)$.

Note that the distance between neighbors is not longer than D. Hence, the set

$$B = \{(\tilde{p}, \hat{q}, \hat{\theta}) | E(\tilde{p}, \hat{q}, \hat{\theta}) \leq E(\tilde{p}(0), \hat{q}(0), \hat{\theta}(0))\} \tag{5.28}$$

is positively invariant, where $\tilde{p} \in D_p$, and $D_p = \{\tilde{p} | \hat{p}_{ij} \in (0, 2D), \forall (i, j) \in \varepsilon\}$. Since \$G(t)\$ is connected for all $t \geq 0$, one gets $\hat{p}_{ij} < 2(N - 1)D$ for all i and j. Since $E(\tilde{p}(t), \hat{q}(t), \hat{\theta}(t)) \leq E(\tilde{p}(0), \hat{q}(0), \hat{\theta}(0))$, we have $\hat{q}^T(t)\hat{q}(t) \leq 2E(\tilde{p}(0), \hat{q}(0), \hat{\theta}(0))$, that is, $\|\hat{q}\| \leq \sqrt{2E(\tilde{p}(0), \hat{q}(0), \hat{\theta}(0))}$. In addition, $\hat{\theta}_i \in (-2\pi, 2\pi), i = 1, \ldots, N$. Thus, the set B is closed and bounded, hence compact. Note that the system (5.1) with control input (5.6) and (5.9) is an autonomous system on the concerned time interval $[t_f, \infty)$.

Then, according to the LaSalle-Krasovskii invariance principle [34], the trajectories of the followers will converge to the invariant set $S = \{(\tilde{p}, \hat{q}, \hat{\theta}) | dE/dt = 0\}$. Clearly, d$E/dt = 0$ if and only if $\hat{q}_i = \hat{q}_j$ and $\dot{\hat{p}}_i = \dot{\hat{p}}_j$ for all $i \in \mathbb{F}$ and $j \in N_i$. Then, we have $\hat{q}_i = \hat{q}_j = \hat{q}_1$ for $i \in N_l, j \in N_i$. With $\hat{q}_1(t) = 0$, one gets $q_i = q_1$ and $q_j - \omega_1 q_1 = 0$. If $j \notin N_l$, one gets $q_j = 0$, which means that agent j stops moving. The state of agent j cannot be maintained, because its neighbor i always follows the moving leader and the distance between agent i and agent j will change to cause the control protocol (5.9) to work. Thus, there is only $j \in N_l$. As mentioned above, followers asymptotically approach a configuration that every follower has one leader neighbor, that is, $N_l = \mathbb{F}$. Discussion about $\dot{\hat{p}}_i = \dot{\hat{p}}_j$ is similar. In a word, $dE/dt = 0$ means $q_i = q_1$ and $\dot{p}_i = \dot{p}_1$ for $i = 2, \ldots, N$.

According to (5.1), $\dot{p}_i = \dot{p}_1$ is equivalent to

$$\begin{cases} \upsilon_i \cos\theta_i - \omega_i l_d \sin\theta_i = \upsilon_1 \cos\theta_1 - \omega_1 l_d \sin\theta_1 \\ \upsilon_i \sin\theta_i + \omega_i l_d \cos\theta_i = \upsilon_1 \sin\theta_1 + \omega_1 l_d \cos\theta_1 \end{cases} \tag{5.29}$$

In terms of $q_i = q_1$ and $l_d > 0$, (5.29) can be explicitly expressed by

$$\begin{cases} \left(\upsilon_1^2 + l_d^2\omega_1^2\right)\sin\theta_i = \left(\upsilon_1^2 + l_d^2\omega_1^2\right)\sin\theta_1 \\ \left(\upsilon_1^2 + l_d^2\omega_1^2\right)\cos\theta_i = \left(\upsilon_1^2 + l_d^2\omega_1^2\right)\cos\theta_1 \end{cases} \tag{5.30}$$

With $\omega_1 \neq 0$, thus we have $\theta_i = \theta_1 \in [0, 2\pi)$. As previously mentioned, the forward speed, the rotational speed, and the heading angle of each follower asymptotically become the same with those of the leader, that is, $\lim_{t\to\infty}|\upsilon_i(t) - \upsilon_1(t)| = 0$, $\lim_{t\to\infty}|\omega_i(t) - \omega_1(t)| = 0$, and $\lim_{t\to\infty}|\theta_i(t) - \theta_1(t)| = 0$, $i \in \mathbb{F}$. Conclusion 3 of Theorem 5.1 is proved.

(4) **Cohesion Formation**

Due to $\dot{q}_1 = 0$ and $q_i = q_1$, one gets $\dot{q}_1 = 0$, $i \in \mathbb{F}$. Here, 0 denotes the zero vector. With $\hat{\theta}_i = \hat{\theta}_j$, $\hat{q}_i = \hat{q}_j$, and $\dot{\hat{p}}_i = \dot{\hat{p}}_j$, $\dot{q}_1 = 0$ can be simplified as

$$\begin{cases} -\sum\limits_{j\in N_i} \nabla_{\hat{p}_{ij}} V\left(\|\hat{p}_{ij}\|\right)\vec{t}_i = 0 \\ -\sum\limits_{j\in N_i} \nabla_{\hat{p}_{ij}} V\left(\|\hat{p}_{ij}\|\right)\vec{n}_i = 0 \end{cases} \tag{5.31}$$

For $\vec{t}_i$ and $\vec{n}_i$ are unit vectors orthogonal to each other, as well as $l_d \neq 0$, (5.31) is equivalent to

$$\sum_{j\in N_i} \nabla_{\hat{p}_{ij}} V\left(\|\hat{p}_{ij}\|\right) = 0,\ i \in \mathbb{F} \tag{5.32}$$

where $N_i = \mathbb{L} \cup \mathbb{F}/\{i\}$. In further, we get

$$\sum_{i\in\mathbb{F}}\sum_{i\in\mathbb{F}} \nabla_{\hat{p}_{ij}} V\left(\|\hat{p}_{ij}\|\right) = 0 \tag{5.33}$$

that is, $\sum_{i\in\mathbb{F}}\sum_{j\in N_i/\{1\}} \nabla_{\hat{p}_{ij}} V\left(\|\hat{p}_{ij}\|\right) + \sum_{j\in N_i} \nabla_{\hat{p}_{i1}} V(\|\hat{p}_{i1}\|) = 0$. Owing to $\sum_{i\in\mathbb{F}}\sum_{j\in N_i/\{1\}} \nabla_{\hat{p}_{ij}} V\left(\|\hat{p}_{ij}\|\right) = 0$, one obtains that $\sum_{j\in N_i} \nabla_{\hat{p}_{i1}} V(\|\hat{p}_{i1}\|) = 0$, that is,

$$\sum_{j\in N_i} \nabla_{\hat{p}_{ij}} V(\|\hat{p}_{i1}\|) = 0 \tag{5.34}$$

where $N_l = \mathbb{F}$ (5.32) and (5.34) mean that the total potential of the system is minimized. Therefore, we get

$$\frac{dV}{dt} = \sum_{i \in \mathbb{F}} \sum_{j \in N_i/\{1\}} \nabla_{\hat{p}_{ij}} V\left(\|\hat{p}_{ij}\|\right) + \sum_{j \in N_i} \nabla_{\hat{p}_{1j}} V\left(\|\hat{p}_{1j}\|\right) = 0 \tag{5.35}$$

Conclusion 4 of Theorem 5.1 is proved.

5.2.3.2 Formation Flocking

Formation control is one of the widely studied topics in multi-robot collaboration research. The primary application of formation control is to utilize the spatial distribution of individual robots, particularly when each robot's sensor has limited capability to gather information. By having each robot's sensor only responsible for collecting environmental data in its immediate vicinity, the system as a whole can comprehensively obtain information about the entire operational area. Formation control research typically covers aspects such as formation establishment, maintenance, and transformation. Currently, there are three main methods for formation control [35]: leader–follower methods [12, 36–38], behavior-based methods [39], and virtual structure methods [40, 41].

According to the above analysis, the whole system stabilizes the inter-agent distances when the total potential of the system is minimized. Thus, any desired rigid formations can be obtained, provided that the potential function reaches its minimal value at the point of the desired inter-agent distance.

Given a desired geometric pattern χ that has N vertices $p_i^d = \left[x_i^d, y_i^d\right]^T$, $i = 1, \ldots, N$. The follower's control protocol protocol is revised by

$$\begin{aligned} a_i(t) =& - \sum_{j \in N_i(t)} \left(\hat{v}_i - \hat{v}_j\right) - \sum_{j \in N_i(t)} \left(\dot{\hat{p}}_i^T - \dot{\hat{p}}_j^T\right)\vec{t}_i \\ &- \sum_{j \in N_i(t)} \nabla_{\tilde{p}_{ij}} V\left(\|\tilde{p}_{ij}\|\right)^T \vec{t}_i \\ b_i(t) =& -l_d \sum_{j \in N_i(t)} \left(\hat{\theta}_i - \hat{\theta}_j\right) - l_d \sum_{j \in N_i(t)} \left(\hat{\omega}_i - \hat{\omega}_j\right) \\ &- \sum_{j \in N_i(t)} \left(\dot{\hat{p}}_i^T - \dot{\hat{p}}_j^T\right)\vec{n}_i - \sum_{j \in N_i(t)} \nabla_{\tilde{p}_{ij}} V\left(\|\tilde{p}_{ij}\|\right)\vec{n}_i \end{aligned} \tag{5.36}$$

where $\|\tilde{p}_{ij}\| = \frac{\|\hat{p}_{ij}\|}{\|p_{ij}^d\|}$, and the potential function $V\left(\|\tilde{p}_{ij}\|\right)$ reaches its minimal value at the point of $\|\tilde{p}_{ij}\| = \sqrt{\frac{\sqrt{b^2+16abD^2}-b}{2a}}$. If $\sqrt{\frac{\sqrt{b^2+16abD^2}-b}{2a}} = 1$, the robotic fish system converges to a table formation exactly coincident with the desired geometric pattern χ. $\sqrt{\frac{\sqrt{b^2+16abD^2}-b}{2a}} > 1$ and $\sqrt{\frac{\sqrt{b^2+16abD^2}-b}{2a}} < 1$ respectively correspond to the expansive formation and contractive formation.

Then, we have the following Corollary 5.1 for the leader–follower formation flocking problem. The proof of Corollary 5.1 is similar to that of Theorem 5.1. Thus, it is omitted here.

Corollary 5.1 (Leader–follower formation flocking) *Consider a system of N agents with dynamics* (5.1)*. The leader and followers are respectively steered by control protocols* (5.6) *and* (5.36)*. Suppose that the initial interaction topology G(0) of the system is a leader–follower connected graph. Then the following statements hold*:

1. *No collision happens between neighbors.*
2. *The connectivity of the interaction topology is preserved at all times.*
3. *The forward speed, the rotational speed, and the heading angle of each follower asymptotically become the same as those of the leader.*
4. *The system approaches a desired geometric configuration χ that minimizes the total potential.*

5.2.4 Experiment Results

5.2.4.1 Robotic Fish

Since the robotic fish shown in Fig. 5.4 takes the carangiform movement mode[42], the thrusting force is commonly generated by means of the undulatory motion of the fish body, which employs the swimming form of a traveling wave described by [42]

$$y_{body}(x,t) = A(x)\sin(kx + \omega t) \tag{5.37}$$

where y_{body} denotes the transverse displacement of the fish body, x denotes the displacement along the main axis, t denotes time, $A(x) = c_1x+c_2x^2$ denotes the wave amplitude, c_1 and c_2 are adjustable parameters, $k = 2\pi/\lambda$ denotes the body wave number, λ denotes the body wave length, $\omega = 2\pi f$ denotes the circular frequency of oscillation, and f denotes the frequency of oscillation.

To simulate the aforementioned traveling wave, the body of the robotic fish is composed of multiple joints connected in sequence [44]. The physical implementation of the proposed traveling body wave is to control each joint $j(j = 1, 2, 3)$ of the robotic fish $i(i = 1, \ldots, N)$ by

$$\psi_i^j = \phi_i^j + A_i^j sin\left(2\pi t f_i^j + \varphi_i^j\right), \qquad j = 1, 2, 3 \tag{5.38}$$

Let subscript j denote the joint number, $\psi_i^j(t)$ the angular position of the jth joint of the robotic fish i, $\phi_i^j(t)$ the angular offset, $A_i^j(t)$ the amplitude, $f_i^j(t)$ the oscillatory

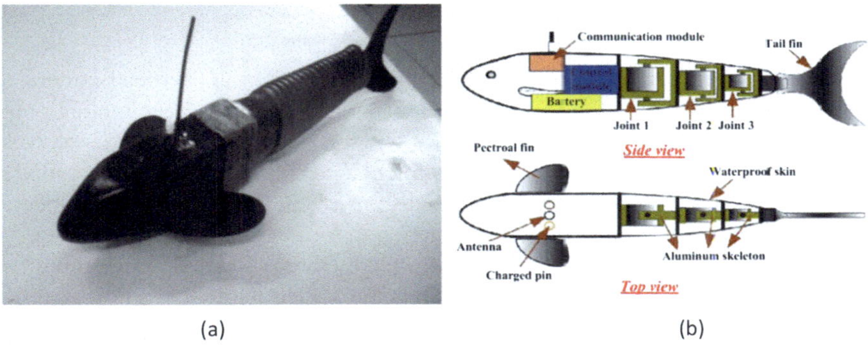

Fig. 5.4 The multi-link biomimetic robotic fish. **a** The prototype. **b** The side view and top view

frequency, and φ_i^j the lag angle. In particular, the lag angles of the robotic fish i satisfy the constraints that $\varphi_i^2 - \varphi_i^1 = 40°$ and $\varphi_i^3 - \varphi_i^1 = 169.3°$, for $i = 1,\ldots,N$.

It should be pointed out that the forward speed $v_i(t)$ of the robotic fish i is related to the oscillatory frequencies $f_i^j(t)$ and the amplitudes $A_i^j(t)$, while the rotational speed $\omega_i(t)$ of the robotic fish i is related to the angular offsets $\phi_i^j(t)$. Based on the proposed control mechanism of the robotic fish, a fuzzy logic controller is developed to build the relationship between A_i^j, f_i^j, ϕ_i^j and ω_i, v_i. Figure 5.5 shows the schematic diagram of the fuzzy logic controller (Fig. 5.6).

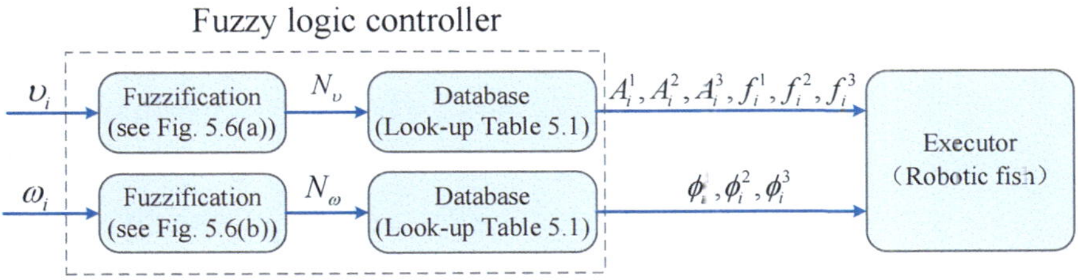

Fig. 5.5 The schematic diagram of the fuzzy logic controller

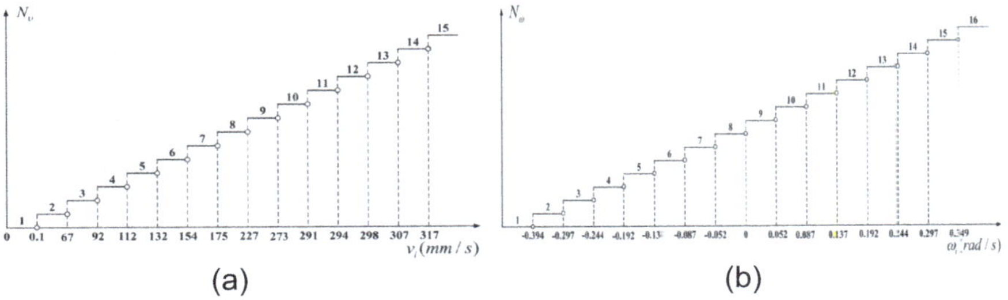

Fig. 5.6 The membership functions. **a** N_υ versus υ_i. **b** N_ω versus ω_i

5.2.4.2 Experiment Environment

As shown in Fig. 5.7, the experiment environment comprises an overhead camera, a host personal computer (PC), an RF communication system, and a swimming pool with dimension of 3 m × 2 m × 0.5 m (length × width × depth). The robotic fish serves as an actuator with limited computation/interaction capability. An overhead camera is utilized to identify and locate each robotic fish. Besides, robotic fish communicate with each other via the host PC as a base station to form the network (Table 5.1).

Two control algorithms are respectively tested by three robotic fish swimming in the surface of the pool. The red robotic fish is the leader, while the yellow robotic fish and the green robotic fish are the followers. The two followers can communicate with each other, and each follower has a leader neighbor. The leader robotic fish swims as the predesigned trajectory under the speed grades $N_v = 8$ and $N_\omega = 12$. The initial speeds of the two followers are zero. The followers adjust their behaviors by using control protocol (5.9) to execute the cohesive flocking task, and apply control protocol (5.36) to adjust their behaviors for formation flocking. During the experiments, a trajectory tracking system embedded into the experiment platform is used to capture the position information and the direction information of each robotic fish in real time. The flowchart of the cohesive experiment is shown as Fig. 5.8. The flowchart of the formation experiment is similar to that of the cohesive experiment, which just needs to replace the control protocol (5.9) by (5.36).

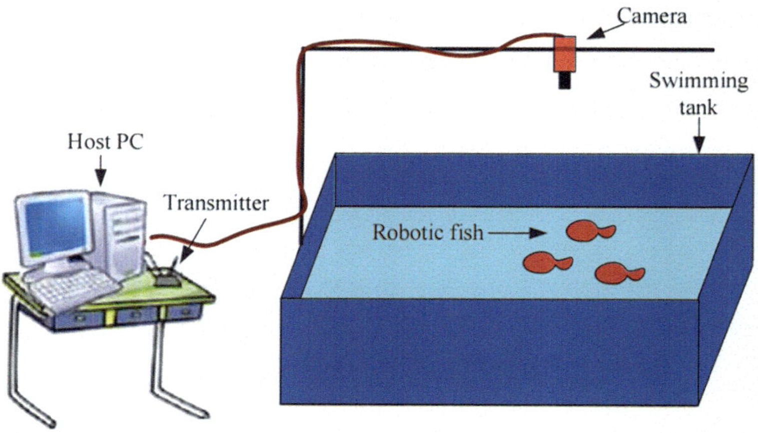

Fig. 5.7 Experiment platform

Table 5.1 Speed grades N_{υ}, N_{ω} of the robotic fish i versus relative control commands $A_i^j(°), f_i^j(Hz)$, $\phi_i^j(°), j = 1, 2, 3$

N_υ	A_i^1	A_i^2	A_i^3	$f_i^1 = f_i^2 = f_i^3$	N_ω	ϕ_i^1	ϕ_i^2	ϕ_i^3
0	0	0	0	0	1	− 54	− 59	− 64
1	2	12	15	0.15	2	− 48	− 53	− 58
2	3	13	18	0.25	3	− 42	− 47	− 52
3	4	14	19	0.266	4	− 30	− 35	− 40
4	5	15	20	0.286	5	− 18	− 23	− 28
5	6	16	21	0.308	6	− 12	− 17	− 22
6	7	17	22	0.354	7	− 6	− 11	− 16
7	8	18	23	0.364	8	0	0	0
8	9	19	24	0.4	9	0	0	0
9	10	20	25	0.44	10	6	11	16
10	11	21	26	0.5	11	12	17	22
11	12	22	27	0.572	12	18	23	28
12	13	23	28	0.666	13	30	35	40
13	14	24	29	0.8	14	42	47	52
14	15	25	30	1	15	48	53	58
15	16	26	31	1.34	16	54	59	64

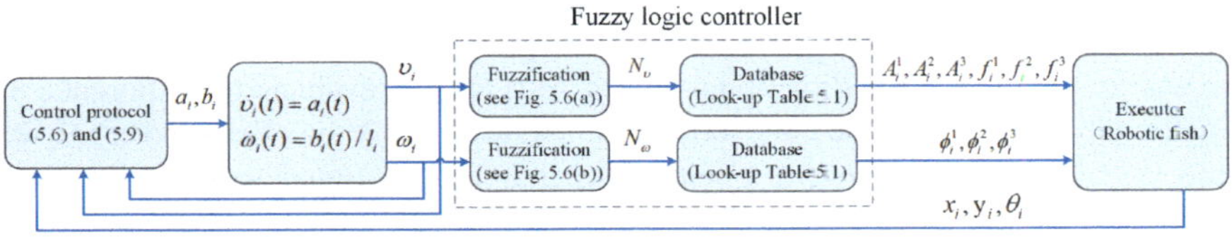

Fig. 5.8 The flow chart of the leader–follower cohesive flocking task implemented by multiple robotic fish

5.2.4.3 Experiment Results

(1) Cohesive Flocking Experiment

Figure 5.9 shows a section of the experimental process, in which the red curve, the yellow curve, and the green curve respectively denote the trajectories of the corresponding robotic fish. Under the action of control protocols (5.6), the two follower robotic fish asymptotically converge to a cohesive formation together with the leader and swim with the same speeds and the same heading angles as the leader. The approximate stable formation is maintained from 13 to 33 s. Experiment one emphasizes that the stable cohesive flocking behaviors of three robotic fish are obtained from a loose and miscellaneous initial condition.

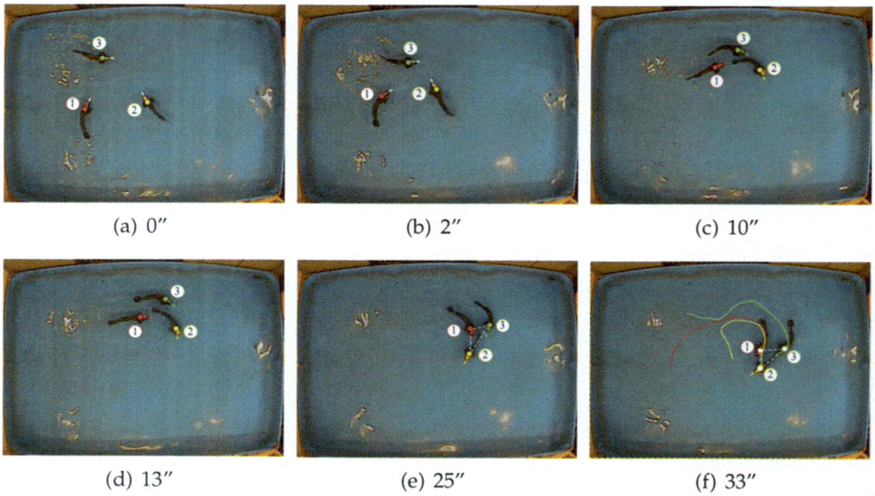

Fig. 5.9 Experiment scenarios of the leader–follower cohesive flocking tasks

Figure 5.10a shows the trajectory of three robotic fish. The green circles represent the initial positions of the robotic fish, while the red squares indicate their positions at $t = 33$ s. The three robotic fish gradually converge and eventually form an aggregated formation. Let $L = \begin{bmatrix} 0 & 1 & 1 \\ 1 & 0 & 1 \\ 1 & 1 & 0 \end{bmatrix}$ and $\theta = [\theta_1, \theta_2, \theta_3]^T$. Define $\Theta = \frac{1}{2}\theta(t)^T L(t)\theta(t)$ as the energy function of the directional angles. Θ is used to evaluate the difference in directional angles among the three robotic fish. Figure 5.10b shows the trend of Θ over time. Θ gradually approaches zero, indicating that the directional angles of the three robotic fish have aligned. Similarly, the time response of the total potential energy is shown in Fig. 5.10c. The total potential energy generally exhibits a downward trend, except for some minor fluctuations, which are believed to be caused by external uncertainties such as water wave disturbances. Eventually, the total potential energy stabilizes at a steady value.

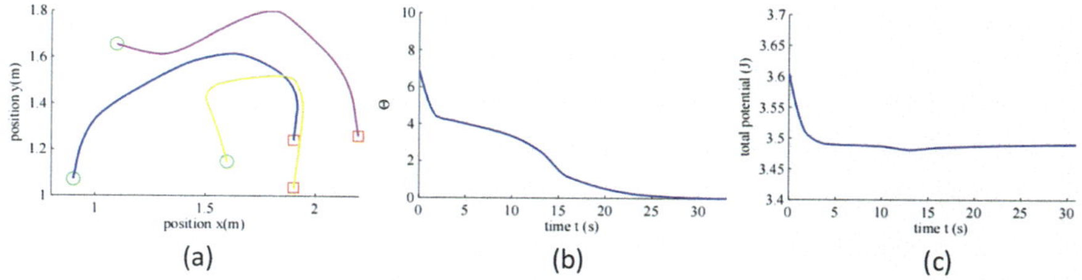

Fig. 5.10 Analysis of cohesive flocking experiment results for the leader–follower robotic fish system, **a** trajectory, **b** the energy function of the direction Angle, **c** the total potential

(2) **Formation Flocking Experiment**

The positions of the vertex of the lambdoid-shaped formation are chosen as $p_1^{d1} = (-0.8, 0)$, $p_2^{d1} = (0, -0.4)$, $p_3^{d1} = (0.8, 0)$. The experiment scenarios of the time response of the attitudes of the three robotic fish are shown in Fig. 5.11, which tells us that three robotic fish asymptotically form and maintain a lambdoid-shaped formation. Therein, in the last scenarios, the red curve, the green curve, and the yellow curve respectively denote the trajectories of the corresponding robotic fish. From 15 to 29 s, a lambdoid-shaped formation can be clearly seen. Different from cohesive flocking experiment, formation flocking experiment focuses on validating that the three robotic fish asymptotically converge to a desired geometric-shaped formation with a consistent swimming velocity. Since $a = 625$, $b = 1$, and $D = 30$, the final stable formation configuration satisfies that $\left\|\hat{p}_{ij}\right\| = 1.5489\left\|p_{ij}^{d1}\right\|$.

Figure 5.12a shows the trajectories of three robotic fish. The green circles represent the initial positions of the robotic fish, while the red squares indicate their positions at $t = 29$ s. The three robotic fish gradually form a V-shaped formation. Figure 5.12b presents the time variation of Θ, which gradually approaches zero, indicating that the heading angles of the three robotic fish become aligned. Similarly, the time response of the total potential energy is shown in Fig. 5.12c.

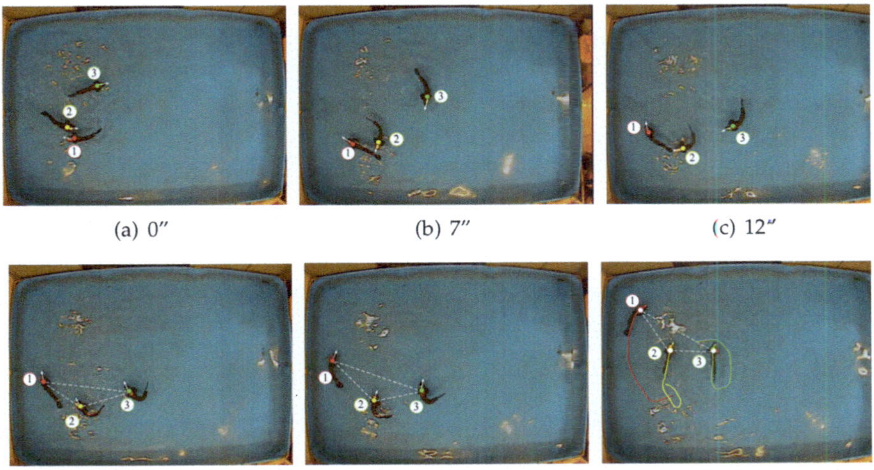

(a) 0″ (b) 7″ (c) 12″

(d) 15″ (e) 19″ (f) 29″

Fig. 5.11 Experiment scenarios of the leader–follower lambdoid-shaped flocking tasks

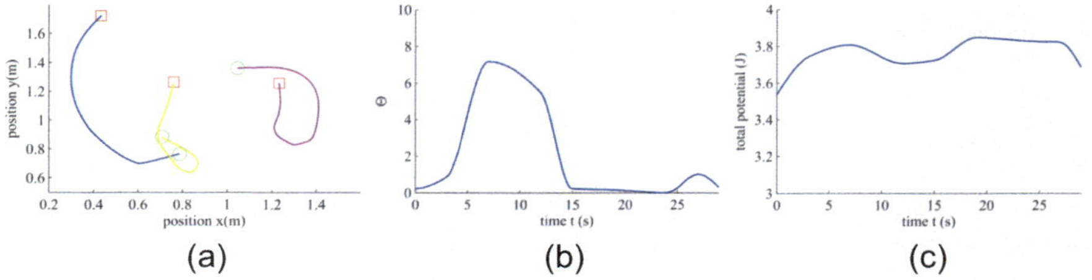

Fig. 5.12 Analysis of cohesive flocking experiment results for the leader–follower lambdoid-shaped robotic fish system, **a** trajectory, **b** the energy function of the direction angle, **c** the total potential

5.2.5 Formation Simulation for Large Group

To better illustrate the proposed approaches for flocking control of a large number of agents, numerical simulations are carried out on Matlab. Furthermore, we also evaluate the performance of the proposed flocking algorithms on a robotic fish simulation platform, involving the proposed kinematic constraint and the proposed control mechanism of the robotic fish.

5.2.5.1 Numerical Simulations

The sample step is 0.01 s and the simulation time is 200 s. Let $S_i(t) = (x_i(t), y_i(t), \theta_i(t), v_i(t), \omega_i(t))$ denote the state vector of the robotic fish i at time t, and the units of the components are respectively m, m, rad, m/s, rad/s. The initial conditions are given by

$$\begin{aligned}
S_1(0) &= (-12.2, -13.09, -0.34, 0.76, -0.29) \\
S_2(0) &= (3.7, -11.78, -1.76, 0.54, 0.7) \\
S_3(0) &= (13.8, -4, 0.6, 5.3, 0.08) \\
S_4(0) &= (4.82, 0.1, -1.1, 7.8, -0.38) \\
S_5(0) &= (-5.8, 4.6, -0.07, 9.34, 0.013) \\
S_6(0) &= (4.97, 19.2, -0.84, 1.3, -0.22) \\
S_7(0) &= (-2.6, 6.1, 0.34, 5.7, -0.275) \\
S_8(0) &= (15.1, 0.6, -0.92, 4.7, -0.34) \\
S_9(0) &= (-4.56, 8.45, -0.12, 0.12, 0.6) \\
S_{10}(0) &= (5.45, -1.42, -0.92, 3.37, 0.31)
\end{aligned} \tag{5.39}$$

(1) **Cohesive Flocking**

Figure 5.13a shows the time response of the ten agents' positions. The green star point is the initial position of each agent, and the rough black circle denotes the final position of each agent. The rough red curve denotes the trajectory of the leader. It is clear to see that ten agents with generic initial conditions asymptotically converge to stable flocking configuration. Figure 5.13b, c tell us that the forward speed, the rotational speed, and the heading angles of the followers asymptotically become the same as those of the leader. Figure 5.13d gives the distances between any two agents. The distances are all asymptotically stable and there is no zero-value distance at all times. Thus, no collision happens during the evolutionary process. Besides, the initial adjacency matrix of the multi-agent system is

$$A_{10}(0) = \begin{bmatrix} 1&0&0&0&0&0&0&0&0&0 \\ 1&0&1&1&1&0&1&1&1&1 \\ 1&1&0&1&1&1&1&1&1&1 \\ 1&1&1&0&1&1&1&1&1&1 \\ 1&1&1&1&0&1&1&1&1&1 \\ 0&0&1&1&1&0&1&1&1&1 \\ 1&1&1&1&1&1&0&1&1&1 \\ 0&1&1&1&1&1&1&0&1&1 \\ 1&1&1&1&1&1&1&1&0&1 \\ 1&1&1&1&1&1&1&1&1&0 \end{bmatrix} \tag{5.40}$$

which indicates that the followers' communication topology is a connected graph but not a complete graph, and only part of the followers have a leader neighbor. We output the adjacency matrix of the whole system in real time, according to which we conclude that the connectivity of the system is preserved at all times. Eventually, the interaction topology of the entire system stabilizes and can be represented by the following adjacency matrix:

$$A_{10}(200) = \begin{bmatrix} 1&0&0&0&0&0&0&0&0&0 \\ 1&0&1&1&1&1&1&1&1&1 \\ 1&1&0&1&1&1&1&1&1&1 \\ 1&1&1&0&1&1&1&1&1&1 \\ 1&1&1&1&0&1&1&1&1&1 \\ 1&1&1&1&1&0&1&1&1&1 \\ 1&1&1&1&1&1&0&1&1&1 \\ 1&1&1&1&1&1&1&0&1&1 \\ 1&1&1&1&1&1&1&1&0&1 \\ 1&1&1&1&1&1&1&1&1&0 \end{bmatrix} \tag{5.41}$$

Finally, the communication topology of the system is fixed to satisfy that $\hat{G}(t)$ is a complete graph and $N_L = \mathbb{F}$. So far, conclusions of Theorem 5.1 are all validated through numerical simulation.

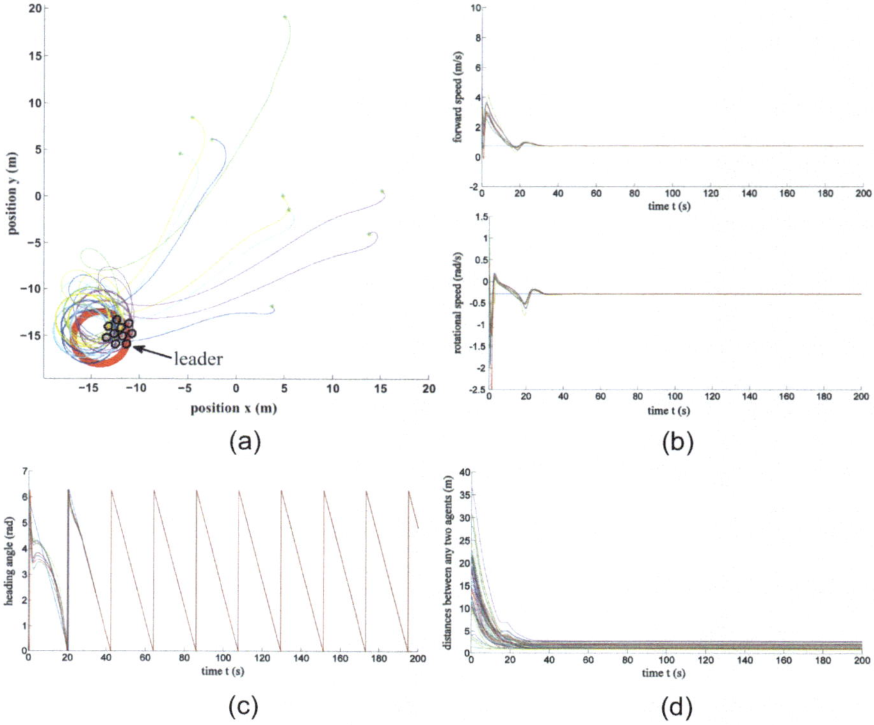

Fig. 5.13 Numerical simulation results of the leader–follower cohesive flocking task. **a** Trajectories of ten agents. **b** Thrusting speeds and rotational speeds with time t. **c** Heading angles with time t. **d** Distances between any two agents with time t

(2) **Formation Flocking**

Two flocking tasks with different formation configuration are involved. One is lambdoid-shaped formation flocking, and the other is ring-shaped formation flocking. Take lambdoid-shaped formation for example, the positions of the vertex are chosen as $p_1^{d2} = (-4, 0)$, $p_2^{d2} = (-3.2, 0.6)$, $p_3^{d2} = (-2.4, 1.2)$, $p_4^{d2} = (-1.6, 1.8)$, $p_5^{d2} = (-0.8, 2.4)$, $p_6^{d2} = (0, 3)$, $p_7^{d2} = (0.8, 2.4)$, $p_8^{d2} = (1.6, 1.8)$, $p_9^{d2} = (2.4, 1.2)$, $p_{10}^{d2} = (3.2, 0.6)$.

With $a = 625$, $b = 1$, and $D = 30$, the geometric constraint to generate the lambdoid-shaped formation is $\|\hat{p}_{ij}\| = 1.5489\left\|p_{ij}^{d2}\right\|$. Similarly, the positions of the vertex of a given ring-shaped formation are chosen as $p_i^{d3} = (R_r * \cos(i * \pi/5), R_r * \sin(i * \pi/5))$, $i = 1, \ldots, 10$, where $R_r > 0$. The geometric constraint to generate the ring-shaped formation is $\|\hat{p}_{ij}\| = 1.5489\left\|p_{ij}^{d3}\right\|$.

The main difference between cohesive flocking and formation flocking refers to the final formation configuration. Hence, for the sake of compact and succinct structure, we only give the time response of the ten agents' position, corresponding to conclusion 4 of Corollary 5.1. Let $R_r = 2m$. Figure 5.14a, b respectively show the trajectory of the ten agents to complete the lambdoid-shaped flocking task and the ring-shaped flocking task.

5.2.5.2 Robotic Fish Platform Simulation

The swimming field is 9 m length and 6 m width. The sample step of the simulation platform is 0.11 s. We pick up the following initial conditions:

$$\begin{aligned}
S_1(0) &= (0.528, 1.416, -2.8623, 0.05, 0.2),\\
S_2(0) &= (-3.88, 0.941, -1.2217, 0, 0),\\
S_3(0) &= (3.166, -2.257, -3.0718, 0, 0),\\
S_4(0) &= (-0.783, 0.582, 1.2566, 0, 0),\\
S_5(0) &= (2.071, -0.910, 2.0071, 0, 0),\\
S_6(0) &= (-0.366, -1.441, -0.8378, 0, 0),\\
S_7(0) &= (-2.006, -1.629, -2.6704, 0, 0),\\
S_8(0) &= (-2.527, 0.08, -1.6232, 0, 0),\\
S_9(0) &= (-2.801, -0.421, -1.7628, 0, 0),\\
S_{10}(0) &= (4.051, -1.649, 1.6057, 0, 0).
\end{aligned} \tag{5.42}$$

Let $R_r = 1$ m. The positions of the vertex of the lambdoid-shaped formation are chosen as $p_1^{d2} = (0, 0)$, $p_2^{d2} = (-0.4, -0.4)$, $p_3^{d2} = (-0.8, -0.8)$, $p_4^{d2} = (-1.2, -1.2)$, $p_5^{d2} = (-1.6, -1.6)$, $p_6^{d2} = (0.4, -0.4)$, $p_7^{d2} = (0.8, -0.8)$, $p_8^{d2} = (1.2, -1.2)$, $p_9^{d2} = (1.6, -1.6)$, $p_{10}^{d2} = (2, -2)$.

The flowchart is shown in Fig. 5.15. It is particularly important to note that the process of disambiguating $v_i(t)$ and $\omega_i(t)$ involves determining the forward speed and rotational speed of the robotic fish under the control commands $A_i^1, A_i^2, A_i^3, f_i^1, f_i^2, f_i^3, \emptyset_i^1, \emptyset_i^2, \emptyset_i^3$. This essentially refers to the minimum value within the speed interval corresponding to the forward speed gear N_v and the minimum value within the speed interval corresponding to the rotational speed gear N_ω. Figure 5.16 presents eight key simulation scenarios of the aggregation and flocking task. The nine followers gradually converge to move at the leader's speed, forming an aggregation flock along with the leader.

By means of the robotic fish simulation platform, the evolutionary process of the robotic fish system consisting of one yellow leader and nine red followers could be recorded. Figures 5.16, 5.17, and 5.18 respectively show eight key snapshots of the simulation scenarios of the cohesive flocking task, the lambdoid-shaped flocking task, and the ring-shaped flocking task. During the three flocking tasks, the nine

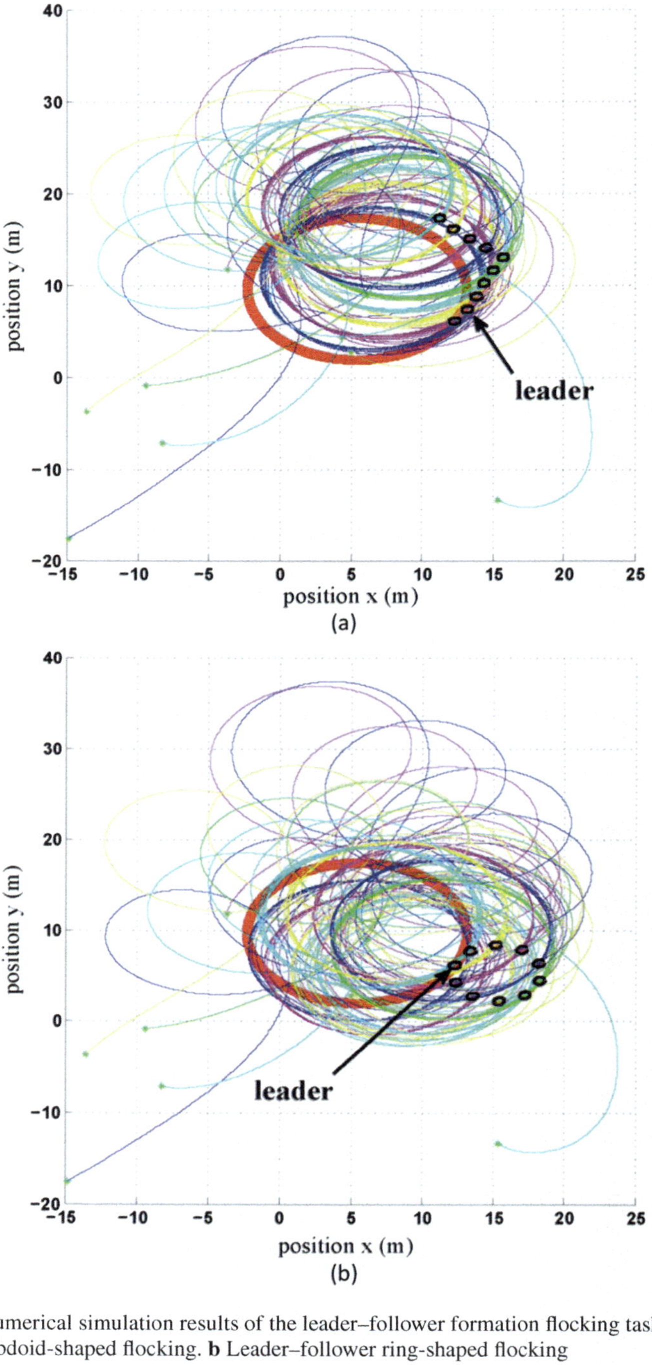

Fig. 5.14 Numerical simulation results of the leader–follower formation flocking task. **a** Leader–follower lambdoid-shaped flocking. **b** Leader–follower ring-shaped flocking

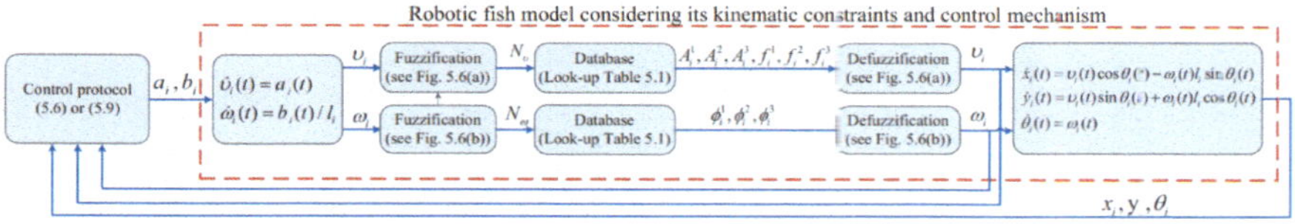

Fig. 5.15 Flowchart of the cohesive flocking control of the leader–follower robotic fish system

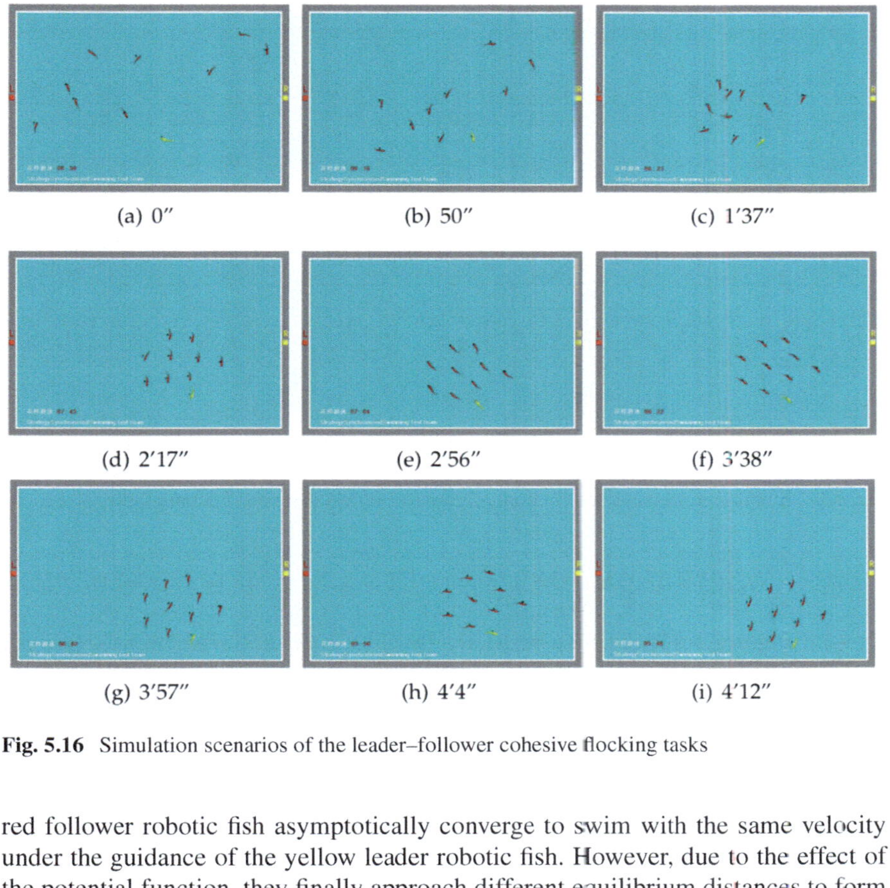

(a) 0″ (b) 50″ (c) 1′37″

(d) 2′17″ (e) 2′56″ (f) 3′38″

(g) 3′57″ (h) 4′4″ (i) 4′12″

Fig. 5.16 Simulation scenarios of the leader–follower cohesive flocking tasks

red follower robotic fish asymptotically converge to swim with the same velocity under the guidance of the yellow leader robotic fish. However, due to the effect of the potential function, they finally approach different equilibrium distances to form a cohesive configuration, a desired lambdoid-shaped configuration, or a ring-shaped configuration together with the yellow leader robotic fish. The evolution processes reveal that stable formations of ten agents under generic initial conditions could be maintained.

The process of formation flocking is similar to that of aggregation flocking, with the control protocol (5.9) being replaced by (5.36). Let $R_r = 1$ m. The coordinates of the vertices in the V-shaped formation are set as follows: $p_1^{d2} = (0, 0), p_2^{d2} =$

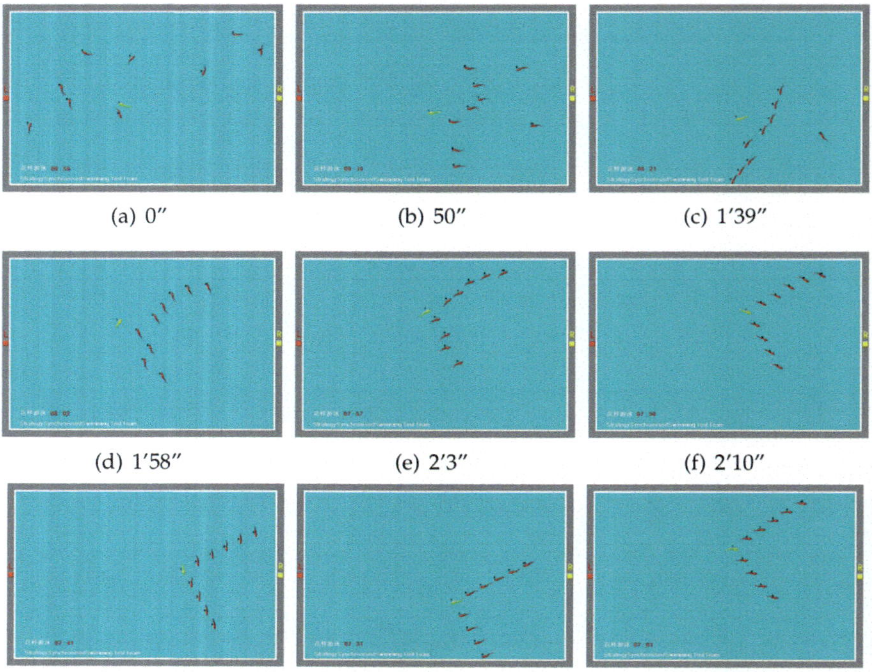

(a) 0″ (b) 50″ (c) 1′39″

(d) 1′58″ (e) 2′3″ (f) 2′10″

(g) 2′19″ (h) 2′29″ (i) 2′57″

Fig. 5.17 Simulation scenarios of the leader–follower lambdoid-shaped flocking tasks

$(-0.4, -0.4)$, $p_3^{d2} = (-0.8, -0.8) p_4^{d2} = (-1.2, -1.2)$, $p_5^{d2} = (-1.6, -1.6)$, $p_6^{d2} = (0.4, -0.4)$, $p_7^{d2} = (0.8, -0.8)$, $p_8^{d2} = (1.2, -1.2)$, $p_9^{d2} = (1.6, -1.6)$, $p_{10}^{d2} = (2, -2)$. Figures 5.17 and 5.18 respectively show eight key simulation scenarios for the V-shaped flocking task and the circular flocking task. The nine followers gradually converge to move at the leader's speed. However, due to the influence of the potential function, the followers eventually stabilize at different equilibrium distances from each other and together with the leader, form the V-shaped or circular flocking formation. The evolution process demonstrates that the stable formation of the ten agents, under general initial conditions, can be consistently maintained.

5.3 Summary

We have investigated the distributed leader–follower cohesive flocking problem and the distributed leader–follower formation flocking problem of multiple robotic fish governed by extended second-order unicycles. Under the guidance of one leader with zero-value external input, the robotic fish system is able to present the desired

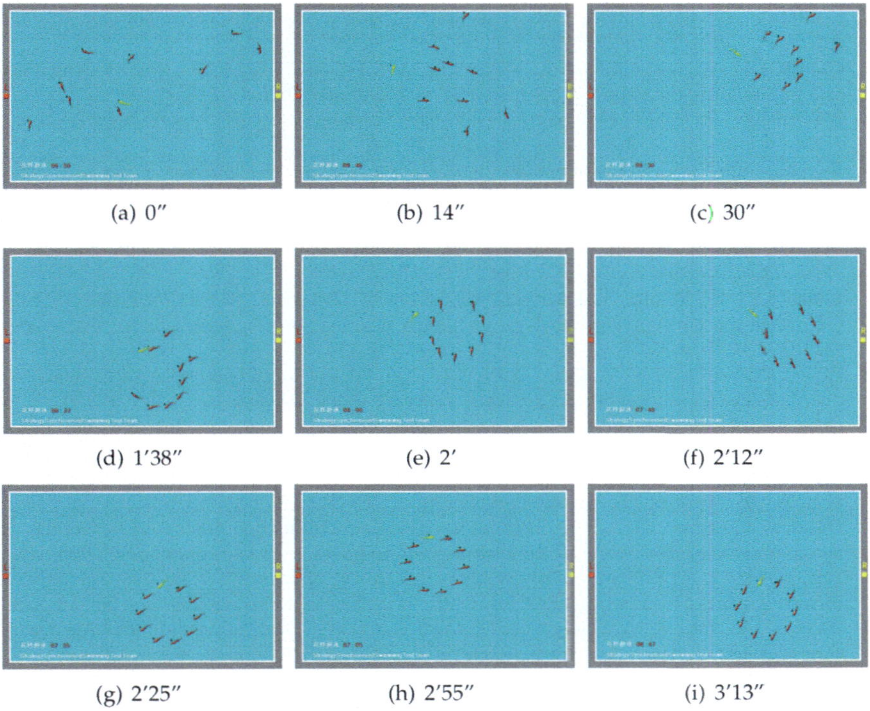

Fig. 5.18 Simulation scenarios of the leader–follower ring-shaped flocking tasks

flocking behaviors over a period of evolution. The stability of the closed-loop system has been analyzed by graph theory and LaSalle-Krasovskii invariance principle. Specially, the collision avoidance between any two robotic fish can be handled by reducing the appointed speed of the leader or increasing the repulsion term of the potential function. The design parameters of the potential function have effect on the formation construction of the robotic fish system. The results of the cohesive flocking experiment and the lambdoid-shaped flocking experiment have verified the feasibility of the proposed flocking algorithms for three robotic fish. Finally, numerical simulations results and the platform simulation results have also been given to further validate the effectiveness of the proposed flocking algorithms for a larger group.

References

1. Rypstra, A. L. (1989). Foraging success of solitary and aggregated spiders: Insights into flocking formation. *Animal Behaviour, 37*, 27.

2. Okubo, A. (1986). Dynamical aspects of animal grouping: Swarms, schools, flocks, and herds. *Advances in Biophysics, 22*, 1–94.
3. Couzin, I. D., Krause, J., James, R., Ruxton, G. D., & Franks, N. R. (2002). Collective memory and spatial sorting in animal groups. *Journal of Theoretical Biology, 218*, 1–11.
4. Olfati-Saber, R. (2006). Flocking for multi-agent dynamic systems: Algorithms and theory. *IEEE Transactions on Automatic Control, 51*, 401–420.
5. Dimarogonas, D. V., & Kyriakopoulos, K. J. (2006). A connection between formation control and flocking behavior in nonholonomic multi-agent systems. In *Proceedings of 2006 IEEE International Conference on Robotics and Automation* (pp. 940–945).
6. Su, H., Wang, X., & Lin, Z. (2009). Flocking for multi-agents with a virtual leader. *IEEE Transactions on Automatic Control, 54*, 293–307.
7. Leonard, N. E., Young, G., Hochgraf, K., Swain, D., Trippe, A., Chen, W., & Marshall, S. (2012). In the dance studio: analysis of human flocking. In *American Control Conference* (pp. 4333–4338).
8. Zhang, G., Fricke, G. K., & Garg, D. P. (2013). Spill detection and perimeter surveillance via distributed swarming agents. *IEEE/ASME Transactions on Mechatronics, 18*, 121–129.
9. Zhan, J., & Li, X. (2013). Flocking of multi-agent systems via model predictive control based on position-only measurements. *IEEE Transactions on Industrial Informatics, 9*, 377–385.
10. Shao, J., Xie, G., & Wang, L. (2007). Leader-following formation control of multiple mobile vehicles. *IET Control Theory and Applications, 1*, 545–552.
11. Desai, J. P., Ostrowski, J. P., & Kumar, V. (2001). Modelling and control of formations of nonholonomic mobile robots. *IEEE Transactions on Robotics and Automation, 17*, 905–908.
12. Das, A., Fierro, R., Kumar, V., Ostrowski, J., Spletzer, J., & Taylor, C. (2002). A vision based formation control framework. *IEEE Transactions on Robotics and Automation, 18*, 813–825.
13. Huang, J., Farritor, S. M., Qadi, A., & Goddard, S. (2006). Localization and follow-the leader control of a heterogeneous group of mobile robots. *IEEE/ASME Transactions on Mechatronics, 11*, 205–215.
14. Ji, M., Muhammad, A., & Egerstedt, M. (2006). Leader-based multi-agent coordination: controllability and optimal control. In *American Control Conference* (pp. 1358–1363).
15. Yang, E., & Gu, D. (2007). Nonlinear formation-keeping and mooring control of multiple autonomous underwater vehicles. *IEEE/ASME Transactions on Mechatronics, 12*, 164–178.
16. Consolini, L., Morbidi, F., Prattichizzo, D., & Tosques, M. (2009). Stabilization of a hierarchical formation of unicycle robots with velocity and curvature constraints. *IEEE Transactions on Robotics, 25*, 1176–1184.
17. Mariottini, G. L., Morbidi, F., Prattichizzo, D., Valk, N. V., Michael, N., Pappas, G., & Daniilidis, K. (2009). Vision-based localization for leader-follower formation control. *IEEE Transactions on Robotics, 25*, 1431–2143.
18. Chen, H., Sun, D., Yang, J., & Chen, J. (2010). Localization for multirobot formations in indoor environment. *IEEE/ASME Transactions on Mechatronics, 15*, 561–574.
19. Ailon, A., & Zohar, I. (2012). Control strategies for driving a group of nonholonomic kinematic mobile robots in formation along a time-parameterized path. *IEEE/ASME Transactions on Mechatronics, 17*, 326–336.
20. Leonard, N. E., & Fiorelli, E. (2001). Virtual leaders, artificial potentials and coordinated control of groups. In *Proceedings of the 40th IEEE Conference on Decision Control* (Vol. 3, pp. 2968–2973).
21. Luo, X., Liu, D., Guan, X., & Li, S. (2012). Flocking in target pursuit for multi-agent systems with particial informed agents. *IET Control Theory and Applications, 6*, 560–569.
22. Su, H., Chen, G., Wang, X., & Lin, Z. (2010). Adaptive flocking with a virtual leader of multiple agents governed by nonlinear dynamics. In *29th Chinese Control Conference* (pp. 5827–5832).
23. Su, H., Wang, X., & Lin, Z. (2009). Flocking for multi-agents with a virtual leader. *IEEE Transactions on Automatic Control, 54*, 293–307.
24. Shi, H., Wang, L., & Chu, T. G. (2006). Virtual leader approach to coordinated control of multiple mobile agents with asymmetric interactions. *Physica D: Nonlinear Phenomena, 213*, 51–65.

25. Yu, W., Chen, G., & Cao, M. (2010). Distributed leader-follower flocking control for multi-agent dynamical systems with time-varying velocities. *Systems and Control Letters, 59*, 543–552.
26. Su, H. (2012). Flocking in multi-agent systems with multiple virtual leaders based only on position measurements. *Communications in Theoretical Physics, 57*, 801–807.
27. Lee, D., Franchi, A., Son, H. I., Ha, C. S., Bulthoff, H. H., & Giordano, P. R. (2013). Semi-autonomous haptic teleoperation control architecture of multiple unmanned aerial vehicles. *IEEE/ASME Transactions on Mechatronics, 18*, 1334–1345.
28. Li, Y., Tang, G., Yang, X., & Wang, P. (2012). A flocking algorithm for multi-agent control with multi-leader following strategy. In *24th Chinese Control and Decision Conference* (pp. 3493–3497).
29. Gu, D., & Wang, Z. (2009). Leader-follower flocking: Algorithms and experiments. *IEEE Transactions on Control Systems Technology, 17*, 1211–1219.
30. Su, H., Wang, X., & Chen, G. (2010). Rendezvous of multiple mobile agents with preserved network connectivity. *Systems and Control Letters, 59*, 313–322.
31. Savkin, A. V., & Teimoori, H. (2009). Decentralized formation flocking and stabilization for networks of unicycles. In *Proceedings of the 48th IEEE Conference on Decision Control and the 28th Chinese Control Conference* (pp. 984–989).
32. Tanner, H. G. (2004). On the controllability of nearest neighbour interconnections. In *43rd IEEE Conference on Decision Control* (Vol. 3, p. 2467).
33. Reynolds, C. W. (1987). Flocks, herds, and schools: A distributed behavioral model. *Computational Graphics, 21*, 25–34.
34. Zavlanos, M. M., Jadbabaie, A., & Pappas, G. J. (2007). Flocking while preserving network connectivity. In *46th IEEE Conference on Decision Control* (pp. 2919–2924).
35. Latombe, J. C. (1991). *Robot motion planning*. Kluwer Academic Publishers.
36. Khalil, H. K. (2002). *Nonlinear systems* (third ed.). Prentice Hall.
37. Su, Z., & Lu, J. (2003). Research on formation control of multiple mobile robots. *Robot., 25*, 88–91.
38. Das, A., Fierro, R., Kumar, V., Ostrowski, J., Spletzer, J., & Taylor, C. (2002). A visionbased formation control framework. *IEEE Transactions on Robotics and Automation, 18*, 813–825.
39. Fredslund, J., & Mataric, M. J. (2002). A general algorithm for robot formations using local sensing and minimal communication. *IEEE Transactions on Robotics and Automation, 18*, 837–846.
40. Bicho, E., & Monteiro, S. (2003). Formation control for multiple mobile robots: A nonlinear attractor dynamics approach. *IEEE International Conference on Intelligent Robots and Systems., 2*, 2016–2022.
41. Chio, T., & Tarn, T. (2003). Rules and control strategies of multi-robot team moving in hierarchical formation. *IEEE International Conference on Robotics and Automation., 2*, 2701–2706.
42. Balch, T., & Arkin, R. C. (1998). Behavior-based formation control for multirobot team. *IEEE Transactions on Robotics and Automation, 14*, 926–939.
43. Lawton, J., Beard, R., & Young, B. (2003). A decentralized approach to formation maneuvers. *IEEE Transactions on Robotics and Automation, 16*, 933–941.
44. Ogren, P., Egerstedt, M., & Hu, X. (2002). A control Lyapunov function approach to multi-agent coordination. *IEEE Transactions on Robotics and Automation, 18*, 847–851.
45. Sfakiotakis, M., Lane, D. M., & Davies, J. B. C. (1999). Review of fish swimming modes for aquatic locomotion. *IEEE Journal of Oceanic Engineering, 24*, 237–252.
46. Barrett, D., Triantafyllou, M., Yue, D. K. P., Grosenbaugh, M. A., & Wolfgang, M. J. (1999). Drag reduction in fish-like locomotion. *Journal of Fluid Mechanics, 392*, 183–212.
47. Yu, J., Tan, M., Wang, S., & Chen, E. (2004). Development of a biomimetic robotic fish and its control algorithm. *IEEE Transactions on Systems, Man, and Cybernetics. Part B, Cybernetics, 34*, 1798–1810.

Chapter 6
Application Scenarios and Countermeasures of Distributed Coordination Control Systems with Variable Leader Speed

6.1 Typical Application Scenarios

6.1.1 Agricultural Monitoring

With the widespread use of agricultural drones, the focus in the industry has gradually expanded from merely pesticide spraying to areas such as agricultural information collection and spectral data analysis. To address the limitations of single-drone operations, drone swarm technology is also gaining attention in the agricultural sector. The SAGA project, which stands for Swarm Robotics for Agricultural Applications, aims to assist farmers in mapping weeds in their fields to improve crop yield. This system, funded by ECHORD++, involves a group of multiple drones working together to monitor a farmland area, using onboard machine vision equipment to accurately locate weeds and create a weed map.

In the drone swarm, the drones exchange information with each other, utilizing the data they collect to prioritize operations in the most weed-dense areas. The algorithm is similar to how honeybees in nature collect nectar from the most flower-dense areas. This path optimization technique helps increase operational efficiency.

Dr. Vito Trainni, coordinator of the SAGA project and researcher at the Institute of Cognitive Sciences and Technologies of the Italian National Research Council (CNR), stated, "Applying swarm robotics to precision agriculture represents a shift in technological paradigms with significant potential impact. As the cost of robotic hardware decreases and the capabilities of robots increase, we will soon be able to achieve automated precision agriculture solutions. This requires individual drones to coordinate as a whole to effectively cover large areas and facilitate information exchange and collaborative work. Drone swarm technology provides a solution to such needs. Miniature robots avoid soil compaction, operate only in the gaps between crops to avoid damaging them, use mechanical rather than chemical weeding methods, and can precisely adapt to different farm sizes. The SAGA project proposes precision

Y. Jia, *Distributed Coordination Control System*,
https://doi.org/10.1007/978-981-96-0365-7_6

agriculture solutions, including novel hardware and precise individual control and swarm intelligence technologies".

The SAGA project utilizes machine vision to identify and locate weeds and employs a drone swarm to collaboratively create weed maps. The Dutch company Avular B.V., specializing in industrial-grade drone monitoring and inspection, provides the drones and machine vision hardware solutions for this project.

SAGA is funded by ECHORD++, a European public project aimed at "bringing robotics research from the lab to the market" by focusing on specific application areas, with precision agriculture as a key area of interest. SAGA is a collaborative research project led by the Institute of Cognitive Sciences and Technologies (ISTC-CNR) of the Italian National Research Council (CNR), which provides expertise on drone and robot swarm applications and coordinates SAGA activities; Wageningen University & Research (WUR) offers expertise in agricultural robotics and precision agriculture; and Avular B. V. implements specific solutions.

Additionally, researchers at the Centre for Distributed Robotics at the University of Minnesota in the United States are investigating distributed solar-powered drones, which will eventually enable low-cost surveying of large agricultural fields. Their roles may include early detection of nitrogen deficiencies, plant diseases, and effective water resource management, as well as cooperative coverage of an area for precise pesticide and herbicide spraying.

6.1.2 Logistics Robot Swarm Transportation

In 2017, Google completed a series of tests organized by the Federal Aviation Administration (FAA) and the National Aeronautics and Space Administration (NASA) to assist in testing the aerial traffic management capabilities of drone systems. During these tests, three Wing drones were assigned tasks of picking up and delivering packages while interacting with two Intel drones and one DJI Inspire drone flying in the same airspace. The tests demonstrated that the Wing traffic management platform could autonomously allocate flight paths for all drones under real outdoor conditions, intelligently update and adapt flight routes, notify remote operators of unexpected route changes, and update airspace warnings and route corrections in response to emergencies like forest fires.

In the future, Google's Project Wing and other companies will operate fleets of thousands of drones to deliver packages and perform other functions. At the same time, these drones will need to achieve intelligent navigation, flying inside or around buildings, and adapting to different weather conditions and other drone missions.

The boom in e-commerce driven by the internet has spurred the development of unmanned systems in the logistics industry. Amazon and JD Logistics have already completed test flights of their logistics drones. The widespread operation of drone technology in logistics will inevitably bring complex management issues. In practical scenarios, with millions of concurrent orders, the allocation algorithms need to support how many robots and drones can avoid each other and whether different robots and drones can collaborate smoothly, all of which require the support of drone coordination technology.

6.1.3 Smart Grids

Distributed coordination control methods were initially proposed and widely applied in the multi-agent systems research field to address problems like formation control, consensus control, flocking, and clustering. In recent years, due to their unique advantages, distributed coordination control methods have gained attention in smart grid research and have been 重点研究领域, including virtual power plant control, economic dispatch, microgrid frequency control, and active distribution network reactive power optimization.

Energy and power systems are crucial industries for national development and are closely related to people's daily lives. Due to the large scale and broad geographical range of power systems, many challenges arise in regulating and controlling these systems. For example, the increasing number of distributed generation (DG) sources, storage devices, and controllable loads connected to the grid via inverters has greatly increased the flexibility of grid control and operation modes. Therefore, researching the stable operation control of microgrids is of significant importance. Additionally, the uncertainty in the output of distributed power sources increases reliance on conventional generation units, severely impacting the integration of renewable energy sources such as wind and solar power.

To effectively reduce the requirements for peak-valley adjustment and frequency regulation in power systems, distributed power sources need to be incorporated into the economic stability control process of the power system to improve operational economy and stability. Traditional centralized control methods involve unified information processing and command regulation by a central microgrid controller, but this approach has poor flexibility and scalability and lower reliability in the event of single-point failures. It cannot effectively adapt to the high penetration of distributed power sources and cannot meet the "plug-and-play" development needs. To overcome the limitations of centralized control, distributed coordination control methods have been applied in power system research.

Studies show that distributed coordination control methods can better meet the stability control and economic dispatch needs of power systems in the context of high penetration of distributed power sources. However, after comparing various distributed algorithms, it has been found that there is still significant research deficiency in non-ideal communication, switching topologies, and coordination attacks and security issues. Future research should strengthen the depth of study in these areas to make algorithms more suitable for practical engineering applications.

6.2 Countermeasures Research

The solutions addressed in this chapter are aimed at multi-agent systems with the following characteristics: The system's agents are divided into two categories. One category is leaders who know the trajectory and destination location for the group, and they travel to the destination at a variable speed along the desired trajectory. The other category is followers who do not need to know the overall goal; each follower has the same communication capability, positioning capability, and consistent decision-making mechanism.

The main goal of the solutions proposed in this chapter is to achieve a coordinated group task where all agents move along a specified trajectory at the same speed as the leader, in the same direction, and maintain a tight formation around the leader. Throughout the process, each agent should be able to avoid collisions with others, and the formation should be robust, meaning that even if it is temporarily disrupted by external disturbances, it can quickly readjust to and maintain a tight formation. Additionally, when agents need to execute collaborative tasks according to a specified geometric formation, this chapter also provides formation control solutions based on swarm strategies.

This chapter presents specific solutions with a focus on robotic fish.

6.2.1 Introduction

Migrating flocking, known as leader–follower flocking, is a special flocking phenomenon that the flocking behaviors of the group are led by one or more experienced leaders. In nature, flocking phenomenon can be widely observed among social animals, in which a group of scattered agents maintain a cohesive or geometric formation based on neighbor information and simple rules. Therein, leaders commonly play an important role in guiding the behaviors of the group.

Gu et al. studied a leader–follower flocking problem concerning multiple leaders, and followers use the position of flocking center to keep their connections [1]. The flocking group proposed by Gu et al. was just able to track a specific trajectory led by group leaders [1], while Yu et al. have given a distributed leader–follower algorithm considering the group consisting of one leader with time-variable speeds [2]. However, the leader involved in Yu's work is still a virtual one [2]. Thus, it is necessary to study a distributed flocking algorithm for the multi-agent system consisting of one real leader with varying speeds. Besides, Savkin et al. studied the formation flocking problem aiming to obtain a geometric formation for a network of unicycles with hard constraints, but they didn't involve any leaders [3]. To this end, another task in this paper is to further extend the designed control algorithm to solve the relative formation flocking problem.

Adaptive strategy has been widely applied to synchronization of complex networks, especially for the cases include uncertain terms [4–8]. For example, Hou et al. given a robust adaptive control approach to solve the consensus problem of multi-agent systems, whose dynamics included the uncertainties and external disturbances [4]. In Demetriou et al.'s work, the adaptation of the consensus gains is used in the disagreement terms of local filters for a sensor network consisting of groups of sensors [5]. Min et al. studied adaptive consensus protocols for non-point, non-linear networked Euler–Lagrange systems with unknown parameters [6]. Su et al. introduced local adaptation strategies for both the weights on the velocity navigational feedback and the velocity coupling strengths that enable all agents to synchronize with the virtual leader [7]. Liu et al. proposed an adaptive protocol to solve the consensus problem of multi-agent systems with high-order nonlinear dynamics by using neural networks to approximate the unknown nonlinear system functions [8]. It thus appears that adaptive control could be an effective method approaching to the unknown varying speed of the leader.

In this chapter, we present a distributed adaptive flocking algorithm to solve the cohesive flocking and geometric-formation flocking of multiple robotic fish consisting of only one leader with varying and bounded speeds. Therein, consensus algorithm is adopted for realizing the velocity alignment, while the problems of connectivity preservation and distance equilibrium are all solved by artificial potential field method [9]. Adaptive strategy is introduced to deal with the varying speeds of the leader. The stability of the closed-loop system is analyzed by means of LaSalle-Krasovskii invariance principle, provided that two initial conditions are satisfied: (1) the interaction network among the followers is an undirected connected graph; (2) there exists at least one follower having a leader neighbor. Furthermore, by adding the information of a desired formation topology to the potential function term, the proposed flocking algorithm can be extended to solve arbitrarily shaped formation flocking problem. Finally, the simulation results are given to verify the effectiveness of the proposed algorithms.

6.2.2 Modeling

The multi-agent system under consideration consists of N robotic fish traveling in a two-dimensional Euclidean space. Robotic fish was modeled as a unicycle during the research on cooperation control of multiple robotic fish [10–12]. Considering that the main propulsive force of the robotic fish comes from the latter part of its body, we make further efforts to draw the swimming robotic fish by a dynamic model named an extended second-order unicycle model, whose geometrical center and mass center don't coincide. Let N denote the set of positive integers, R the set of real numbers, and R^+ the set of positive real numbers. As shown in Fig. 6.1, the robotic fish is modeled by the following kinematic equation

$$\dot{x}_i(t) = \upsilon_i(t)\cos\theta_i(t) - \omega_i(t)l_i\sin\theta_i(t)$$
$$\dot{y}_i(t) = \upsilon_i(t)\sin\theta_i(t) + \omega_i(t)l_i\cos\theta_i(t)$$

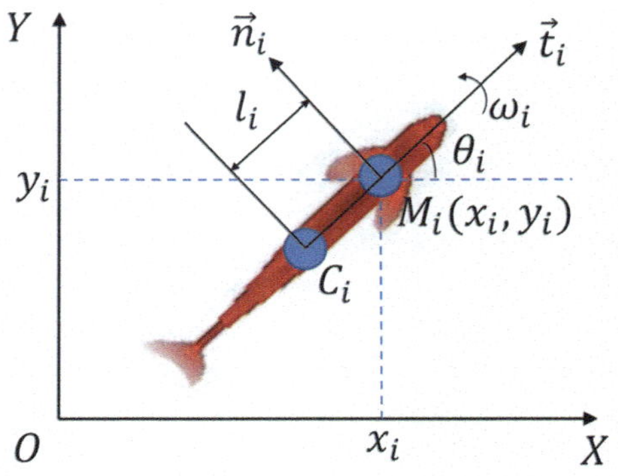

Fig. 6.1 The simplified model of robotic fish

$$\begin{aligned}\dot{\theta}_i(t) &= \omega_i(t)\\ \dot{\upsilon}_i(t) &= a_i(t)\\ \dot{\omega}_i(t) &= b_i(t)/l_i\end{aligned} \tag{6.1}$$

where $p_i(t) = [x_i(t), y_i(t)]^T \in \mathbb{R}^2$ is the position vector of agent i at time t, $\theta_i(t) \in \mathbb{R}$ the heading angle of agent i at time t, $\upsilon_i(t) \in \mathbb{R}$ the thrusting speed of agent i at time t, $\omega_i(t) \in \mathbb{R}$ the rotational speed of agent i at time t, $l_i \in \mathbb{R}^+$ the distance between the geometrical center C_i and the mass center M_i of agent i at time t, $\vartheta_i(t) = \omega_i(t)l_i \in \mathbb{R}$ the tangential speed of agent i at time t, $a_i(t) \in \mathbb{R}$ the thrusting acceleration of agent i at time t, and $b_i(t) \in \mathbb{R}$ the rotational acceleration of agent i at time t. Here, $\theta_i(t) \in [0, 2\pi)$ is measured from the x-axis in the anticlockwise rotation. In this paper, we do not consider the individual difference during the theoretical analysis. Thus, it should be noted that $l_i = l_d, i = 1, \ldots, N$, where $l_d > 0$ is a constant.

We consider the robotic fish system with only one leader. If an agent has external control input, we call it a leader; else, we call it a follower. Without loss of generality, let leader set be $\mathbb{L} = \{1\}$ and follower set be $\mathbb{F} = \{2, \ldots, N\}$. Each agent only interacts with its neighbor due to its limited interaction capability. We suppose that interconnection with the leader is unidirectional, while interconnection with the follower is bidirectional [13]. Let N_i (t) denote the neighbor set of the follower $i \in \mathbb{F}$ at time t, and the initial neighbor set of the follower i is defined as

$$N_i(0) = \{j | \left\| p_i(0) - p_j(0) \right\| < D, j = 1, \ldots, N, j \neq i\} \tag{6.2}$$

where $D > 0$ is a constant. $\| \cdot \|$ represents the Euclidean norm, $D > 0$ is the interaction radius.

The interaction network $G(t)$ is a dynamic directed graph consisting of a vertex set $\nu = \{1, \cdots, N\}$ indexed by robotic fish and a time-varying edge set $\varepsilon(t) = \{(i, j)|(i, j) \in \nu \times \nu, j \in N_i(t)\}$. Therein, the followers' interaction network $\hat{G}(t)$ with vertex set F and edge set $\hat{\varepsilon}(t) = \{(i, j)|(i, j) \in \hat{\nu} \times \hat{\nu}, j \in N_i(t)\}$ is an undirected graph. The matrix $A(t) = [a_{ij}(t)]_{N\times N}$ with definition

$$a_{ij}(t) = \begin{cases} i = 1, \quad j = 1 \\ 1, i \in F, j \in N_i(t) \\ 0, \quad otherwise \end{cases} \tag{6.3}$$

represents the coupling configuration of the network $G(t)$. Let c_{ij} denote the coupling strengths between node i and node j. Define the matrix of the weighted coupling configuration of the network $G(t)$ as follows:

$$B(t) = \left[b_{ij}(t)\right]_{N \times N} = \begin{bmatrix} c_{11}(t)a_{11}(t) & c_{12}(t)a_{12}(t) & \cdots & c_{1N}(t)a_{1N}(t) \\ c_{21}(t)a_{21}(t) & c_{22}(t)a_{22}(t) & \cdots & c_{2N}(t)a_{2N}(t) \\ \vdots & \vdots & \ddots & \vdots \\ c_{N1}(t)a_{N1}(t) & c_{N1}(t)a_{N1}(t) & \cdots & c_{NN}(t)a_{NN}(t) \end{bmatrix} \tag{6.4}$$

where $b_{ij}(t) \in \mathbb{R}, c_{ij}(t) \in \mathbb{R}^+, i, j \in v$. The Laplacian matrix $L_{N-1}(t) = [l_{ij}(t)]_{(N-1)\times(N-1)}$ of the graph $\hat{G}(t)$ is given by

$$l_{ij}(t) = \begin{cases} -c_{ij}(t)a_{ij}(t), i \neq j \\ \sum\limits_{k=2, k\neq i}^{N} c_{ik}(t)a_{ik}(t), i = j \end{cases} \tag{6.5}$$

For $\hat{G}(t)$ is an undirected graph, the Laplacian matrix $L_{N-1}(t)$ is symmetric and positive semi-definite. Figure 6.2 gives a sample of the interaction network for the robotic fish system.

Define that the interaction network $G(t)$ switches at $t_p, p = 1, 2, \ldots$. Thus, $G(t)$ is a fixed graph in each nonempty, bounded, and contiguous time-interval $\left[t_r, t_{r+1}\right)$, where $r = 0, 1, \ldots$ and $t_0 = 0$. Given that $\hat{G}(0)$ is an undirected connected graph, and there exists at least one follower having a leader neighbor at the initial time

Fig. 6.2 An example of the interaction network

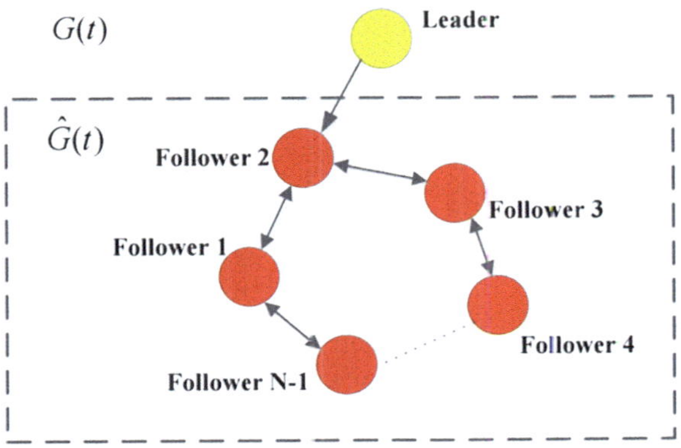

$t_0 = 0$. In order to preserve the connectivity of the interaction network $G(t)$, the hysteresis adding new edges to the network is introduced [14, 15], such that

(1) if $(i,j) \in \varepsilon(t^-)$ and $\|\hat{p}_i(t) - \hat{p}_j(t)\| < 2D$, where $\hat{p}_i(t) = p_i(t) - p_i^*(t)$ and $p_i^*(t) = \int_0^t w_{i1}(t) H_i^T(t) q_1 d\tau (i = 1, \ldots, N) \, (i = 1, \ldots, N)$, then $(i,j) \in \varepsilon(t)$, for $t > 0$;
(2) if $(i,j) \notin \varepsilon(t^-)$ and $\|p_i(t) - p_j(t)\| < D$, then $(i,j) \in \varepsilon(t)$, for $t > 0$.

6.2.3 *Distributed Coordination Control Algorithm Design and Stability Analysis*

6.2.3.1 Algorithm Design

Let $q_i(t) = [\upsilon_i(t), \vartheta_i(t)]^T$ be the velocity vector of agent i, the model (6.1) can be rewritten in matrix form as

$$
\begin{aligned}
\dot{p}_i(t) &= H_i(t)^T q_i(t) \\
\dot{q}_i(t) &= u_i(t)
\end{aligned}
\tag{6.6}
$$

where $H_i(t) = \begin{bmatrix} \cos\theta_i(t) & \sin\theta_i(t) \\ -\sin\theta_i(t) & \cos\theta_i(t) \end{bmatrix}$, $u_i(t) = [a_i(t), b_i(t)]^T$, $i = 1, \ldots, N$, the total potential V of the system is

$$
V = \sum_{i \in \mathbb{F}} \sum_{j \in N_i} V((\|\hat{p}_{ij}\|)) + \sum_{j \in N_l} V(\|\hat{p}_{1j}\|) \tag{6.7}
$$

Assume that motion is synchronous and that no time delays. The external control input of the leader is given by

$$
u_i(t) = f(t) \tag{6.8}
$$

where $f(t) = [f_1(t), f_2(t)]^T \in \mathbb{R}^2$ is bounded, who satisfies that

$$
\lim_{t \to \infty} f_1(t) = 0 \tag{6.9}
$$

$$
\lim_{t \to \infty} f_2(t) = 0 \tag{6.10}
$$

In addition, the control input of the follower i $(i = 2, \ldots, N)$ is designed by

$$
a_i(t) = -\sum_{j \in N_i(t)} c_{ij} \left(\hat{v}_i(t) - \hat{v}_j(t) \right)
$$

$$
\begin{aligned}
& - \sum_{j \in N_i(t)} c_{ij}(t)\dot{\hat{p}}_{ij}(t)^T \vec{t}_i - \sum_{j \in N_i(t)} \nabla_{\hat{p}_{ij}} V\left(\|\hat{p}_{ij}\|\right)\vec{t}_i(t) \\
b_i(t) = & - \sum_{j \in N_i(t)} c_{ij}(t) l_d\left(\hat{\theta}_i(t) - \hat{\theta}_j(t)\right) \\
& - \sum_{j \in N_i(t)} c_{ij}(t) l_d\left(\hat{\omega}_i(t) - \hat{\omega}_j(t)\right) \\
& - \sum_{j \in N_i(t)} \dot{\hat{p}}_{ij}(t)^T \vec{t}_i - \sum_{j \in N_i(t)} c_{ij}(t)\nabla_{\hat{p}_{ij}} V\left(\|\hat{p}_{ij}\|\right)\vec{t}_i \\
\dot{c}_{ij}(t) = & \; k_{ij}\left(\hat{q}_i(t) - \hat{q}_j(t)\right)^T\left(\hat{q}_i(t) - \hat{q}_j(t)\right) + k_{ij}\dot{\hat{p}}_{ij}(t)^T\dot{\hat{p}}_{ij}(t)
\end{aligned}
\tag{6.11}
$$

where $\hat{v}_i(t) = \upsilon_i(t) - a_{i1}(t)\upsilon_1(t)$, $\hat{\omega}_i(t) = \omega_i(t) - a_{i1}(t)\omega_1(t)$, $\hat{\theta}_i(t) = \theta_i(t) - a_{i1}(t)\theta_1(t)$, $\hat{p}_{ij}(t) = \hat{p}_i(t) - \hat{p}_j(t)$, $c_{ij} > 0$, k_{ij} is the weight of the adaptive parameter $c_{ij}(t)$. Here, $p_i^* = \int_0^t a_{i1} H_i(t)^T q_1(t) d\tau$. Thus, we have $\dot{\hat{p}}_i(t) = \dot{p}_i(t) - \dot{p}_i^* = H_i(t)^T q_i(t) - a_{i1} H_i(t)^T q_1(t)$, where $\hat{q}_i(t) = q_i(t) - a_{i1}(t)q_1(t)$. Besides, $\vec{t}_i(t) = \begin{bmatrix} \cos\theta_i(t) \\ \sin\theta_i(t) \end{bmatrix}$ and $\vec{n}_i(t) = \begin{bmatrix} -\sin\theta_i(t) \\ \cos\theta_i(t) \end{bmatrix}$ are two unit vectors orthogonal to each other, and $\nabla_{\hat{p}_{ij}(t)} V\left(\|\hat{p}_{ij}(t)\|\right)$ is the gradient of an artificial potential function $V\left(\|\hat{p}_{ij}(t)\|\right)$ with the same definition as Definition 5.1.

6.2.3.2 Stability Analysis

In this chapter, the potential function $V\left(\hat{p}_{ij}\right)$ between individuals i and j is related to their differences in posture, including position and orientation angles. Then, we have Theorem 6.1 to solve the leader–follower cohesive flocking problem.

Theorem 6.1 (Leader–follower cohesive flocking) *Consider a system of N agents with dynamics (6.1). The leader and the followers are respectively steered by control protocols (6.8) and (6.11). Suppose that the initial interaction network $G(0)$ among the followers is an undirected connected graph, and there exists at least one follower having a leader neighbor at the initial time. Then the following statements hold:*

1. *The connectivity of the interaction network is preserved at all times, that is, $\varepsilon(t_r) \subseteq \varepsilon(t_{r+1})$.*
2. *The thrusting speed, the rotational speed, and the heading angle of each follower asymptotically become the same as those of the leader, that is, $\lim_{t\to\infty}\left(\upsilon_i(t) - \upsilon_j(t)\right) = 0$, $\lim_{t\to\infty}\left(\omega_i(t) - \omega_j(t)\right) = 0$, and $\lim_{t\to\infty}\left(\theta_i(t) - \theta_j(t)\right) = 0$, where $i, j \in \nu$, $i \neq j$.*
3. *The system approaches a cohesive configuration that minimizes the total potential, that is, $\sum_{j\in N_i} \nabla_{\widehat{p_{ij}(t)}} V\left(\|\widehat{p_{ij}}(t)\|\right) = 0$, and $\sum_{j\in N_l} \nabla_{\widehat{p_{1j}(t)}} V\left(\|\widehat{p_{1j}}(t)\|\right) = 0$.*

Proof Consider the system (6.1) with control input (6.8) and (6.11) on time interval $\left[t_r, t_{r+1}\right)$, where the interaction network of the robotic fish system is fixed.

Define that $\tilde{P}(t) = [\hat{p}_{11}(t)^T, \ldots, \hat{p}_{1N}(t)^T, \cdots, \hat{p}_{N1}(t)^T, \cdots, \hat{p}_{NN}(t)^T]^T$, $\hat{q}(t) = [\hat{q}_2(t)^T, \ldots, \hat{q}_N(t)^T]^T$, and $\hat{\theta}(t) = [\hat{\theta}_2(t), \ldots, \hat{\theta}_N(t)]^T$. Let $N_l(t) = \{i | a_{i1}(t) = 1, i \in \mathbb{F}\}$ denote the set of followers who have one leader neighbor on time interval $[t_r, t_{r+1})$. Consider the following energy function as the common Lyapunov function

$$E\big(\tilde{P}(t), \hat{q}(t), \hat{\theta}(t)\big) = E_1(t) + E_2(t) + E_3(t) + E_4(t) \tag{6.12}$$

where

$$E_1(t) = \frac{1}{2} V(t) \tag{6.13}$$

$$E_2(t) = \frac{1}{2} \sum_{i \in \mathbb{F}} \hat{q}_i(t)^T \hat{q}_i(t) \tag{6.14}$$

$$E_3(t) = \frac{1}{2} \Theta(t) \tag{6.15}$$

$$E_4(t) = \frac{1}{4} \sum_{i \in \mathbb{F}} \sum_{j \in N_i(t)} \frac{(c_{ij}(t) - m)^2}{k_{ij}} \tag{6.16}$$

Here, $V(t) = \sum_{i \in \mathbb{F}} \sum_{j \in N_i(t)} V(\hat{p}_{ij}) + \sum_{j \in N_l(t)} V(\hat{p}_{1j}(t))$, and $\Theta(t) = l_d^2 \sum_{i \in \mathbb{F}} \sum_{j \in N_i(t)} c_{ij}(t) \hat{\theta}_i(t)) (\hat{\theta}_i(t) - \hat{\theta}_j(t))$.

Suppose that the interaction network $G(t)$ switches at t_p, $p = 1, 2, \ldots$. $G(t)$ is a fixed graph in each nonempty, bounded, and contiguous time-interval $[t_r, t_{r+1})$, $r = 0, 1, \ldots$. Here, $t_0 = 0$. Then the derivative of $E(\tilde{p}, \hat{q}, \hat{\theta})$ w.r.t. the time $t \in [t_r, t_{r+1})$ is

$$\frac{dE}{dt} = \frac{dE_1}{dt} + \frac{dE_2}{dt} + \frac{dE_3}{dt} + \frac{dE_4}{dt} \tag{6.17}$$

We will respectively express the four parts on the right side of Eq. (6.17) in the following paragraphs.

Firstly, we have that

$$\begin{aligned}
\frac{dE_1}{dt} &= \frac{1}{2} \sum_{i \in \mathbb{F}} \sum_{j \in N_i(t)} \dot{\hat{p}}_{ij}(t)^T \nabla_{\hat{p}_{ij}(t)} V\big(\|\hat{p}_{ij}(t)\|\big) \\
&\quad + \frac{1}{2} \sum_{j \in N_l} \dot{\hat{p}}_{1j}(t)^T \nabla_{\hat{p}_{1j}(t)} V\big(\|\hat{p}_{1j}(t)\|\big) \\
&= \sum_{i \in \mathbb{F}} \dot{\hat{p}}_{ij}(t)^T \sum_{j \in N_i/\{1\}} \nabla_{\hat{p}_{ij}(t)} V\big(\|\hat{p}_{ij}(t)\|\big) \\
&\quad + \frac{1}{2} \sum_{j \in N_i(t)} \dot{\hat{p}}_{i1}(t)^T \nabla_{\hat{p}_{i1}(t)} V(\|\hat{p}_{i1}(t)\|)
\end{aligned}$$

$$\frac{1}{2}\sum_{j\in N_l}\dot{p}_{1j}(t)^T\nabla_{\hat{p}_{1j}(t)}V\left(\|\hat{p}_{1j}(t)\|\right) \tag{6.18}$$

Due to $\dot{\hat{p}}_{ij}(t)^T = -\dot{\hat{p}}_{ji}(t)^T$ and the symmetric nature of $V\left(\|\hat{p}_{ij}(t)\|\right)$, one gets

$$\begin{aligned}
&\frac{1}{2}\sum_{j\in N_l}\dot{\hat{p}}_{1j}(t)^T\nabla_{\hat{p}_{1j}(t)}V\left(\|\hat{p}_{1j}(t)\|\right)\\
&=\frac{1}{2}\sum_{j\in N_l}\dot{\hat{p}}_{j1}(t)^T\nabla_{\hat{p}_{j1}(t)}V\left(\|\hat{p}_{j1}(t)\|\right)\\
&=\frac{1}{2}\sum_{j\in N_l}\dot{\hat{p}}_{i1}(t)^T\nabla_{\hat{p}_{i1}(t)}V(\|\hat{p}_{i1}(t)\|)
\end{aligned} \tag{6.19}$$

With $\dot{\hat{p}}_{i1}(t)^T = \dot{\hat{p}}_i(t)^T - \dot{\hat{p}}_1(t)^T = \dot{\hat{p}}_i(t)^T$, we have

$$\begin{aligned}
\frac{dE_1}{dt} &= \sum_{i\in\mathbb{F}}\dot{\hat{p}}_i(t)^T\sum_{j\in N_l/\{1\}}\nabla_{\hat{p}_{ij}(t)}V\left(\|\hat{p}_{ij}(t)\|\right)\\
&\quad+\sum_{j\in N_l}\dot{\hat{p}}_{i1}(t)^T\nabla_{\hat{p}_{i1}(t)}V(\|\hat{p}_{i1}(t)\|)\\
&=\sum_{i\in\mathbb{F}}\dot{\hat{p}}_i(t)^T\sum_{j\in N_l/\{1\}}\nabla_{\hat{p}_{ij}(t)}V\left(\|\hat{p}_{ij}(t)\|\right)\\
&\quad+\sum_{j\in N_l}\dot{\hat{p}}_i(t)^T\nabla_{\hat{p}_{i1}(t)}V(\|\hat{p}_{i1}(t)\|)\\
&=\sum_{i\in\mathbb{F}}\dot{\hat{p}}_i(t)^T\sum_{j\in N_i}\nabla_{\hat{p}_{ij}(t)}V\left(\|\hat{p}_{ij}(t)\|\right)
\end{aligned} \tag{6.20}$$

Secondly, since $\hat{\theta}_1(t) = 0$, $\Theta(t)$ can be rewritten by

$$\Theta(t) = l_d^2\sum_{i\in\mathbb{F}}\sum_{j\in N_i/\{1\}}c_{ij}(t)\hat{\theta}_i(t)\left(\hat{\theta}_i(t)-\hat{\theta}_j(t)\right) + l_d^2\sum_{i\in N_l}c_{i1}(t)\hat{\theta}_i^2(t) \tag{6.21}$$

Furthermore, we have

$$\begin{aligned}
\frac{dE_2}{dt} &= l_d^2\sum_{i\in\mathbb{F}}\sum_{j\in N_i/\{1\}}c_{ij}(t)\dot{\hat{\theta}}_i(t)\left(\hat{\theta}_i(t)-\hat{\theta}_j(t)\right) + l_d^2\sum_{i\in N_l}c_{i1}(t)\dot{\hat{\theta}}_i(t)\hat{\theta}_i(t)\\
&= l_d^2\sum_{i\in\mathbb{F}}\sum_{j\in N_i}c_{ij}(t)\dot{\hat{\theta}}_i(t)\left(\hat{\theta}_i(t)-\hat{\theta}_j(t)\right)\\
&= l_d^2\sum_{i\in\mathbb{F}}\sum_{j\in N_i}c_{ij}(t)\hat{\omega}_i(t)\left(\hat{\theta}_i(t)-\hat{\theta}_j(t)\right)
\end{aligned} \tag{6.22}$$

Thirdly, let $q_i^*(t) = [0, l_d\theta_i(t)]^T$, $\hat{q}_i^*(t) = q_i^*(t) - a_{i1}(t)q_1^*(t) = [0, l_d\hat{\theta}_i(t)]^T$, $i = 1, \ldots, N$. The control protocol (6.11) can be rewritten as

$$\begin{aligned}\dot{q}_i(t) = &-\sum_{j\in N_i} c_{ij}(t)\Big(\hat{q}_i^*(t) - \hat{q}_j^*(t)\Big) - \sum_{j\in N_i} c_{ij}(t)\big(\hat{q}_i(t) - \hat{q}_j(t)\big)\\ &- H_i(t)\sum_{j\in N_i} c_{ij}(t)\dot{\hat{p}}_{ij}(t) - H_i(t)\sum_{j\in N_i} \nabla_{\hat{p}_{ij}(t)}V\left(\left\|\hat{p}_{ij}(t)\right\|\right)\\ \dot{c}_{ij}(t) = &\ k_{ij}\big(\hat{q}_i(t) - \hat{q}_j(t)\big)^T\big(\hat{q}_i(t) - \hat{q}_j(t)\big)\\ &+ k_{ij}\big(\dot{\hat{p}}_i(t) - \dot{\hat{p}}_j(T)\big)^T\big(\dot{\hat{p}}_i(T) - \dot{\hat{p}}_j(t)\big)\end{aligned} \tag{6.23}$$

Due to $\dot{\hat{q}}_i(t) = \dot{q}_i(t) - a_{i1}\dot{q}_1(t) = \dot{q}_i(t) - a_{i1}f(t)$ and $\dot{\hat{p}}_i(t)^T = \dot{q}_i(t)^T H_i(t)$, we have

$$\begin{aligned}\frac{dE_3}{dt} &= \sum_{i\in\mathbb{F}} \hat{q}_i(t)^T\dot{\hat{q}}_i(t)\\ &= \sum_{i\in\mathbb{F}} \hat{q}_i(t)^T(\dot{q}_i(t) - a_{i1}f(t))\\ &= \sum_{i\in\mathbb{F}} \hat{q}_i(t)^T(\dot{q}_i(t) - a_{i1}f(t))\\ &= -\sum_{i\in\mathbb{F}} \hat{q}_i(t)^T\left(-\sum_{j\in N_i} c_{ij}(t)\Big(\hat{q}_i^*(t) - \hat{q}_j^*(t)\Big)\right)\\ &\quad -\sum_{j\in N_i} c_{ij}(t)\big(\hat{q}_i(t) - \hat{q}_j(t)\big)\\ &\quad - H_i(t)\sum_{j\in N_i} c_{ij}(t)\dot{\hat{p}}_{ij}(t) - H_i(t)\\ &\quad \sum_{j\in N_i} \nabla_{\hat{p}_{ij}(t)}V\left(\left\|\hat{p}_{ij}(t)\right\|\right) - a_{i1}f(t))\\ &= -\sum_{i\in\mathbb{F}} \hat{q}_i(t)^T\sum_{j\in N_i} c_{ij}(t)\Big(\hat{q}_i^*(t) - \hat{q}_j^*(t)\Big)\\ &\quad -\sum_{i\in\mathbb{F}} \hat{q}_i(t)^T\sum_{j\in N_i} c_{ij}(t)\big(\hat{q}_i(t) - \hat{q}_j(t)\big)\\ &\quad -\sum_{i\in\mathbb{F}} \hat{q}_i(t)^T H_i(t)\sum_{j\in N_i} c_{ij}(t)\dot{\hat{p}}_{ij}(t)\\ &\quad -\sum_{i\in\mathbb{F}} \hat{q}_i(t)^T H_i(t)\sum_{j\in N_i} \nabla_{\hat{p}_{ij}(t)}V\left(\left\|\hat{p}_{ij}(t)\right\|\right)\\ &\quad -\sum_{i\in\mathbb{F}} \hat{q}_i(t)^T a_{i1}f(t)\end{aligned}$$

$$
\begin{aligned}
&= l_d^2 \sum_{i\in\mathbb{F}} \sum_{j\in N_i} c_{ij}(t)\hat{\omega}_i(t)\left(\hat{\theta}_i(t) - \hat{\theta}_j(t)\right) \\
&\quad - \sum_{i\in\mathbb{F}} \sum_{j\in N_i} c_{ij}(t)\hat{q}_i(t)^T\left(\hat{q}_i(t) - \hat{q}_j(t)\right) \\
&\sum_{i\in\mathbb{F}} \sum_{j\in N_i} c_{ij}(t)\dot{\hat{p}}_i(t)(\dot{\hat{p}}_i(t) - \dot{\hat{p}}_j(t)) \\
&\quad - \sum_{i\in\mathbb{F}} \dot{\hat{p}}_i(t)^T \sum_{j\in N_i} \nabla_{\hat{p}_{ij}(t)} V\left(\left\|\hat{p}_{ij}(t)\right\|\right) \\
&\quad - \sum_{i\in\mathbb{F}} \hat{q}_i(t)^T a_{i1} f(t)
\end{aligned}
\tag{6.24}
$$

Finally, we have

$$
\begin{aligned}
\frac{dE_4}{dt} &= \frac{1}{2} \sum_{i\in\mathbb{F}} \sum_{j\in N_i} \frac{\left(c_{ij}(t) - m\right)\dot{c}_{ij}(t)}{k_{ij}} \\
&= \frac{1}{2} \sum_{i\in\mathbb{F}} \sum_{j\in N_i} \left(c_{ij}(t) - m\right)(\hat{q}_i(t) - \hat{q}_j(t))^T\left(\hat{q}_i(t) - \hat{q}_j(t)\right)) \\
&\quad + \frac{1}{2} \sum_{i\in\mathbb{F}} \sum_{j\in N_i} \left(c_{ij}(t) - m\right)(\dot{\hat{p}}_i(t) - \dot{\hat{p}}_j(t))^T(\dot{\hat{p}}_i(t) - \dot{\hat{p}}_j(t)) \\
&= \sum_{i\in\mathbb{F}} \sum_{j\in N_i} \left(c_{ij}(t) - m\right)\hat{q}_i(t)^T\left(\hat{q}_i(t) - \hat{q}_j(t)\right) \\
&\quad + \sum_{i\in\mathbb{F}} \sum_{j\in N_i} \left(c_{ij}(t) - m\right)\dot{\hat{p}}_i(t)^T(\dot{\hat{p}}_i(t) - \dot{\hat{p}}_j(t))
\end{aligned}
\tag{6.25}
$$

Thus, $\frac{dE}{dt}$ can be simplified by

$$
\begin{aligned}
\frac{dE}{dt} &= -m \sum_{i\in\mathbb{F}} \hat{q}_i(t)^T \sum_{j\in N_i} \left(\hat{q}_i(t) - \hat{q}_j(t)\right) \\
&\quad - m \sum_{i\in\mathbb{F}} \dot{\hat{p}}_i(t)^T \sum_{j\in N_i} \left(c_{ij}(t) - m\right)\left(\dot{\hat{p}}_i(t) - \dot{\hat{p}}_j(t)\right) \\
&\quad - \sum_{i\in\mathbb{F}} \hat{q}_i(t)^T a_{i1} f(t) \\
&= -m \sum_{i\in\mathbb{F}} \hat{q}_i(t)^T \sum_{j\in N_i} \left(\hat{q}_i(t) - \hat{q}_j(t)\right) \\
&\quad - m \sum_{i\in\mathbb{F}} \dot{\hat{p}}_i(t)^T \sum_{j\in N_i} \left(c_{ij}(t) - m\right)\left(\dot{\hat{p}}_i(t) - \dot{\hat{p}}_j(t)\right) \\
&\quad - \sum_{i\in\mathbb{F}} \hat{q}_i(t)^T f(t)
\end{aligned}
\tag{6.26}
$$

Owing to $\hat{q}_1 = 0$, one gets

$$\begin{aligned}
&- m\sum_{i\in\mathbb{F}} \hat{q}_i(t)^T \sum_{j\in N_i}\big(\hat{q}_i(t) - \hat{q}_j(t)\big) \\
&= -m\sum_{i\in\mathbb{F}} \hat{q}_i(t)^T \sum_{j\in N_i/\{1\}} \big(\hat{q}_i(t) - \hat{q}_j(t)\big) - m\sum_{i\in\mathbb{F}} \hat{q}_i(t)^T\hat{q}_i(t) \\
&= -m\hat{q}(t)^T(L_{N-1}\otimes I_2)\hat{q}(t) - m\sum_{i\in\mathbb{F}} \hat{q}_i(t)^T\hat{q}_i(t) \le 0
\end{aligned} \tag{6.27}$$

where I_2 denotes the 2×2 identity matrix, and L_{N-1} is symmetric and positive semi-definite. Let 0 denote the zero vector. Besides, $-m\sum_{i\in\mathbb{F}}\hat{q}_i(t)^T\sum_{j\in N_i}\big(\hat{q}_i(t)-\hat{q}_j(t)\big) = 0$ if and only if $\hat{q}_i(t) = \hat{q}_j(t)$ for all $i\in\mathbb{F}$ and $j\in N_i$. Similarly, due to $\dot{\hat{p}}_1(t) = 0$, we have

$$\begin{aligned}
&- m\sum_{i\in\mathbb{F}} \dot{\hat{p}}_i(t)^T \sum_{j\in N_i}\big(\dot{\hat{p}}_i(t) - \dot{\hat{p}}_j(t)\big) \\
&= -m\dot{\hat{p}}(t)^T(L_{N-1}\otimes I_2)\dot{\hat{p}}(t) - m\sum_{j\in N_i} \dot{\hat{p}}_i(t)^T\dot{\hat{p}}_i(t) \le 0
\end{aligned} \tag{6.28}$$

where $\hat{p}(t) = \big[\hat{p}_2(t)^T,\ldots,\hat{p}_N(t)^T\big]^T$, and $-m\sum_{i\in\mathbb{F}}\dot{\hat{p}}_i(t)^T\sum_{j\in N_i}\big(\dot{\hat{p}}_i(t)-\dot{\hat{p}}_j(t)\big) = 0$ if and only if $\dot{\hat{p}}_i(t) = \dot{\hat{p}}_j(t)$ for all $i\in\mathbb{F}$ and $j\in N_i$.

If $\hat{q}_i(t) = \hat{q}_j(t)$ and $\dot{\hat{p}}_i(t) = \dot{\hat{p}}_j(t)$, we have

$$\begin{aligned}
\frac{dE}{dt} &= -m\hat{q}(t)^T(L_{N-1}\otimes I_2)\hat{q}(t) - m\sum_{i\in N_l} \hat{q}_i(t)^T\hat{q}_i(t) \\
&\quad - m\dot{\hat{p}}(t)^T(L_{N-1}\otimes I_2)\dot{\hat{p}}(t) \\
&\quad - m\sum_{i\in N_l}\dot{\hat{p}}_i(t)^T\dot{\hat{p}}_i(t) - \sum_{i\in N_l}\hat{q}_i(t)^T f(t) = 0
\end{aligned} \tag{6.29}$$

If $\hat{q}_i(t) \neq \hat{q}_j(t)$ or $\dot{\hat{p}}_i(t) \neq \dot{\hat{p}}_j(t)$, then we have $-m\sum_{i\in\mathbb{F}}\hat{q}_i(t)^T\sum_{j\in N_i}\big(\hat{q}_i(t)-\hat{q}_j(t)\big) < 0$ or $-m\sum_{i\in\mathbb{F}}\dot{\hat{p}}_i(t)^T\sum_{j\in N_i}\big(\dot{\hat{p}}_i(t)-\dot{\hat{p}}_j(t)\big) < 0$. Since $f(t)$ is bounded, as long as the positive constant m is sufficiently large, then we have

$$\begin{aligned}
dE/dt &= -m\hat{q}(t)^T(L_{N-1}\otimes I_2)\hat{q}(t) - m\sum_{i\in N_l} \hat{q}_i(t)^T\hat{q}_i(t) \\
&\quad - m\dot{\hat{p}}(t)^T(L_{N-1}\otimes I_2)\dot{\hat{p}}(t) \\
&\quad - m\sum_{i\in\mathbb{F}}\dot{\hat{p}}_i(t)^T\dot{\hat{p}}_i(t) - \sum_{i\in\mathbb{F}}\hat{q}_i(t)^T f(t) < 0
\end{aligned} \tag{6.30}$$

Therefore, we have $dE/dt \le 0$.

The initial potential energy of the system is finite, and the initial speeds and initial heading angles of all agents are also finite. Thus, the initial energy $E(\tilde{p}(0), \hat{q}(0), \hat{\theta}(0))$ of the system is finite, and the supremum of E is obviously its initial value $E(\tilde{p}(0), \hat{q}(0), \hat{\theta}(0))$. Then, the potential energy V of the system is finite, and the potential energy $V(\|\hat{p}_{ij}(t)\|)$ between agent i and j is finite.

If for some $(i, j) \in \varepsilon$, $\|\hat{p}_{ij}\| \to 2D$, then according to rule (2) of Definition 5.1, one gets $V(\|\hat{p}_{ij}\|) \to \infty$, which violates the conclusion that $V(\|\hat{p}_{ij}\|)$ remains finite. It follows that $\|\hat{p}_{ij}\| < 2D$ for all $(i, j) \in \varepsilon$ and $t \in [t_r, t_{r+1})$. Hence, whenever there is a link between two agents, the link is never lost during each time-interval $[t_r, t_{r+1})$, that is, $\varepsilon(t_r) \subseteq \varepsilon(t_{r+1})$. Note that when $\|\hat{p}_{ij}\| < 2D$, one gets $\|p_{ij}\| < R = 2D + \|p^*_{ij}\|$. It is obvious that $R > D$, which ensures that if an edge $(i, j) < \varepsilon$ is added to ε, the associated potential $V(\hat{p}_{ij})$ is bounded and hence, so is the new potential V. Consequently, the connectivity of the communication network can be preserved all the time. Conclusion 1 of Theorem 6.1 is proved.

Assume that there are $m_r \in N$ new links being added to the interaction network at switching time $t_r (r = 1, 2, \cdots)$. We have supposed that the initial interaction network should satisfy the following two conditions: (1) the interaction network among the followers is an undirected connected graph; (2) there exists at least one follower having a leader neighbor. Let G_1 denote the initial interaction network of the system, and G_c denote the set of all graphs meeting the proposed two conditions on the vertices. The proposed control algorithm can guarantee that the sequence of switching topologies G_{r+1} within $[t_r, t_{r+1})$ consists of such graphs satisfying $G_{r+1} \in G_c$. The number of the vertices is finite, thus G_c is a finite set. Assume that there are at most $M \in N$ new links that can be added to the initial interaction network G_1. Clearly, $0 < m_r \le M$ and $r \le M$. Therefore, the number of switching times of the system is finite, and the interaction network $G(t)$ eventually becomes fixed. Suppose that the last switching time is t_f, the following discussions are restricted on the time interval $[t_f, \infty)$.

Note that the distance between neighbors is not longer than D. Hence, the set

$$B = \{(\tilde{p}, \hat{q}, \hat{\theta}) | E(\tilde{p}(t), \hat{q}(t), \hat{\theta}(t)) \le E(\tilde{p}(0), \hat{q}(0), \hat{\theta}(0))\} \tag{6.31}$$

is positively invariant, where $\tilde{p} \in D_p, D_p = \{\tilde{p} | \|\hat{p}_{ij}\| \in (0, 2D), \forall (i, j) \in \varepsilon\}$. Since G(t) is connected for all t $\ge$ 0, one gets $\|\hat{p}_{ij}\| <$ 2(N $-$ 1)D for all i and j. Since $E(\tilde{p}(t), \hat{q}(t), \hat{\theta}(t)) \le E(\tilde{p}(0), \hat{q}(0), \hat{\theta}(0))$, we have $\hat{q}^T \hat{q} \le 2\sqrt{E(\tilde{p}(0), \hat{q}(0), \hat{\theta}(0))}$, that is, $\|\hat{q}\| \le 2E(\tilde{p}(0), \hat{q}(0), \hat{\theta}(0))$. In addition, $\hat{\theta}_i \in (-2\pi, 2\pi), i = 1, \cdots, N$. Thus, the set B is closed and bounded, hence compact. Note that the system (6.1) with control input (6.8) and (6.11) is an autonomous system on the concerned time interval $[t_f, \infty)$.

Then, according to the LaSalle-Krasovskii invariance principle [16], the trajectories of the followers will converge to the invariant set $S = \{(\tilde{p}, \hat{q}, \hat{\theta}) | dE/dt = 0\}$. Clearly, $dE/dt = 0$ if and only if $\hat{q}_i(t) = \hat{q}_j(t)$ and $\dot{\hat{p}}_i(t) = \dot{\hat{p}}_j(t)$ for all $i \in \mathbb{F}$ and $j \in N_i$. Then, we have $\hat{q}_i(t) = \hat{q}_j(t) = \hat{q}_1(t)$ for $i \in N_l, j \in N_i$. With $\hat{q}_1(t) = 0$, one gets $q_i(t) = q_1(t)$ and $q_j(t) - a_{ji} q_1(t) = 0$. If $j \notin N_l$, one gets $q_j(t) = 0$, which means that agent j stops moving. The state of agent j cannot be maintained, because its neighbor i always follows the moving leader and the distance between agent i and agent j will change to cause the control protocol (6.11) to work. Thus, there is only $j \in N_l$. As mentioned above, followers asymptotically approach a configuration that every follower has one leader neighbor, that is, $N_l = \mathbb{F}$. Discussion about $\dot{\hat{p}}_i(t) = \dot{\hat{p}}_j(t)$ is similar. In a word, $dE/dt = 0$ means $q_i(t) = q_1(t)$ and $\dot{p}_i(t) = \dot{p}_1(t)$ for $i = 2, \cdots, N$.

According to (6.1), $\dot{p}_i(t) = \dot{p}_1(t)$ is equivalent to

$$\begin{cases} \upsilon_i(t)\cos\theta_i(t) - \omega_i(t)l_d\sin\theta_i(t) = \upsilon_1(t)\cos\theta_1(t) - \omega_1(t)l_d\sin\theta_1(t) \\ \upsilon_i(t)\sin\theta_i(t) + \omega_i(t)l_d\cos\theta_i(t) = \upsilon_1(t)\sin\theta_1(t) + \omega_1(t)l_d\cos\theta_1(t) \end{cases} \tag{6.32}$$

In terms of $q_i(t) = q_1(t)$ and $l_d > 0$, (6.32) can be explicitly expressed by

$$\begin{cases} \left(\upsilon_1^2(t) + l_d^2\omega_1^2(t)\right)\sin\theta_i(t) = \left(\upsilon_1^2(t) + l_d^2\omega_1^2(t)\right)\sin\theta_1(t) \\ \left(\upsilon_1^2(t) + l_d^2\omega_1^2(t)\right)\cos\theta_i(t) = \left(\upsilon_1^2(t) + l_d^2\omega_1^2(t)\right)\cos\theta_1(t) \end{cases} \tag{6.33}$$

With $q_1(t) \neq 0$, thus we have $\theta_i(t) = \theta_1(t) \in [0, 2\pi)$. As previously mentioned, the thrusting speed, the rotational speed, and the heading angle of each follower asymptotically become the same with those of the leader, that is,

$$\lim_{t\to\infty}|\upsilon_i(t) - \upsilon_1(t)| = 0,\ \lim_{t\to\infty}|\omega_i(t) - \omega_1(t)| = 0,\ \lim_{t\to\infty}|\theta_i(t) - \theta_1(t)| = 0,\ i \in \mathbb{F},$$

where $i, j \in \mathrm{v}$, $i \neq j$. Conclusion 2 of Theorem 6.1 is proved.

Due to $\lim_{t\to\infty} f_1(t) = 0$ and $\lim_{t\to\infty} f_2(t) = 0$, that is, $\lim_{t\to\infty}\dot{q}_1(t) = 0$, one gets $\dot{q}_i(t) = \dot{q}_1(t) = 0$ in steady state, $i \in \mathbb{F}$. With $\hat{\theta}_i(t) = \hat{\theta}_j(t)$, $\hat{q}_i(t) = \hat{q}_j(t)$, and $\dot{\hat{p}}_i(t) = \dot{\hat{p}}_j(t)$, $\dot{q}_1(t) = 0$ can be simplified as

$$\begin{cases} -\sum\limits_{j\in N_i}\nabla_{\hat{p}_{ij}(t)}V\left(\left\|\hat{p}_{ij}(t)\right\|\right)^T\vec{t_i}(t) = 0 \\ -\sum\limits_{j\in N_i}\nabla_{\hat{p}_{ij}(t)}V\left(\left\|\hat{p}_{ij}(t)\right\|\right)^T\vec{n_i}(t) = 0 \end{cases} \tag{6.34}$$

For $\vec{t_i}(t) = \begin{bmatrix}\cos\theta_i(t)\\ \sin\theta_i(t)\end{bmatrix}$ and $\vec{n_i}(t) = \begin{bmatrix}-\sin\theta_i(t)\\ \cos\theta_i(t)\end{bmatrix}$ are unit vectors orthogonal to each other, as well as $l_d \neq 0$, (6.34) is equivalent to

$$\sum_{i\in\mathbb{F}}\sum_{j\in N_i}\nabla_{\hat{p}_{ij}(t)}V\left(\left\|\hat{p}_{ij}(t)\right\|\right) = 0,\ i \in \mathbb{F} \tag{6.35}$$

where $N_i = \mathbb{L} \cup \mathbb{F}/\{i\}$. In further, we get

$$\sum_{i\in\mathbb{F}}\sum_{j\in N_i}\nabla_{\hat{p}_{ij}(t)}V\left(\left\|\hat{p}_{ij}(t)\right\|\right) = 0 \tag{6.36}$$

that is, $\sum_{i\in\mathbb{F}}\sum_{j\in N_i/\{1\}}\nabla_{\hat{p}_{ij}(t)}V\left(\left\|\hat{p}_{ij}(t)\right\|\right)+\sum_{j\in N_l}\nabla_{\hat{p}_{i1}(t)}V(\|\hat{p}_{i1}(t)\|)=0$. Owing to $\sum_{i\in\mathbb{F}}\sum_{j\in N_i/\{1\}}\nabla_{\hat{p}_{ij}(t)}V\left(\left\|\hat{p}_{ij}(t)\right\|\right)=0$, one obtains that $\sum_{j\in N_l}\nabla_{\hat{p}_{i1}(t)}V(\|\hat{p}_{i1}(t)\|)=0$, that is,

$$\sum_{j\in N_l}\nabla_{\hat{p}_{1j}(t)}V\left(\left\|\hat{p}_{1j}(t)\right\|\right)=0 \tag{6.37}$$

where $N_l=\mathbb{F}$. Equations (6.35) and (6.37) mean that the total potential of the system is minimized. Thus, conclusion 3 of Theorem 6.1 is proved.

According to the above analysis, the whole system stabilizes the inter-agent distances when the total potential of the system is minimized. Thus, any desired rigid formations can be obtained, provided that the potential function reaches its minimal value at the point of the desired inter-agent distance. Given a desired geometric pattern χ that has N vertices $p_i^d=\left[x_i^d,y_i^d\right]^T$, $i=1,\ldots,N$. The follower's control protocol is revised by

$$\begin{aligned}
a_i(t)=&-\sum_{j\in N_i(t)}c_{ij}(t)\left(\hat{v}_i(t)-\hat{v}_j(t)\right)-\sum_{j\in N_i(t)}c_{ij}(t)\dot{\hat{p}}_{ij}(t)^T\vec{t}_i(t)\\
&-\sum_{j\in N_i(t)}\nabla_{\tilde{p}_{ij(t)}}V\left(\left\|\tilde{p}_{ij}(t)\right\|\right)^T\vec{t}_i(t)\\
b_i(t)=&-\sum_{j\in N_i(t)}c_{ij}(t)l_d\left(\hat{\theta}_i(t)-\hat{\theta}_j(t)\right)-\sum_{j\in N_i(t)}c_{ij}(t)l_d\left(\hat{\omega}_i(t)-\hat{\omega}_j(t)\right)\\
&-\sum_{j\in N_i(t)}c_{ij}(t)\dot{\hat{p}}_{ij}(t)^T\vec{n}_i(t)-\sum_{j\in N_i(t)}\nabla_{\tilde{p}_{ij(t)}}V\left(\left\|\tilde{p}_{ij}(t)\right\|\right)^T\vec{n}_i(t)\\
\dot{c}_{ij}(t)=&\,k_{ij}\left(\hat{q}_i(t)-\hat{q}_j(t)\right)^T\left(\hat{q}_i(t)-\hat{q}_j(t)\right)+k_{ij}\dot{\hat{p}}_{ij}(t)^T\dot{\hat{p}}_{ij}(t)
\end{aligned} \tag{6.38}$$

where $\left\|\tilde{p}_{ij}(t)\right\|=\frac{\left\|\hat{p}_{ij}(t)\right\|}{\left\|p_{ij}^d(t)\right\|}$, and the potential function $V(\|\tilde{p}_{ij}(t)\|)$在$\|\tilde{p}_{ij}(t)\|=1$ is required to reach its minimal value at the point of $\left\|\tilde{p}_{ij}(t)\right\|=1$. Then, we have the following Corollary 6.1.

Corollary 6.1 (Leader–Follower Formation Flocking) *Consider a system of N agents with dynamics (6.1). The leader and the followers are respectively steered by control protocols (6.8) and (6.38). Suppose that the initial interaction network G(0) among the followers is an undirected connected graph, and there exists at least one follower having a leader neighbor at the initial time. Then the following statements hold*:

1. *The connectivity of the interaction network is preserved at all times, that is,* $D=30$.
2. *The thrusting speed, the rotational speed, and the heading angle of each follower asymptotically become the same as those of the leader, that is,* $\lim_{t\to\infty}\left(\upsilon_i(t)-\upsilon_j(t)\right)=0$, $\lim_{t\to\infty}\left(\omega_i(t)-\omega_j(t)\right)=0$, *and* $\lim_{t\to\infty}\left(\theta_i(t)-\theta_j(t)\right)=0$, *where* $i,j\in\nu$, $i\neq j$.

3. *The system approaches a desired geometric configuration* χ *that minimizes the total potential, that is,* $\sum_{j\in N_i} \nabla_{\widetilde{p_{ij}}(t)} V\left(\left\|\widetilde{p_{ij}}(t)\right\|\right) = 0$, *and* $\sum_{j\in N_l} \nabla_{\widetilde{p_{1j}}(t)} V\left(\left\|\widetilde{p_{1j}}(t)\right\|\right) = 0$.

The proof of Corollary 6.1 is similar to that of Theorem 6.1. Thus, it is omitted here.

6.2.4 Simulation Validation and Discussion

In this section, we will show the simulation results of the proposed theoretical results. According to Definition 5.1, we design the following potential function

$$V\left(\left\|\hat{p}_{ij}\right\|\right) = \frac{b}{\left\|\hat{p}_{ij}\right\|^2} - aln\left(4D^2 - \left\|\hat{p}_{ij}\right\|^2\right) + c \tag{6.39}$$

where a and b are positive constants, c is constant. Therein, the first part is the repulsive potential term, and the second part is the attractive potential term, and c is just used to guarantee the potential function positive at all times. Then, the gradient of the potential function is obtained by

$$\nabla_{\hat{p}_{ij}} V\left(\left\|\hat{p}_{ij}\right\|\right) = 2\hat{p}_{ij}\left(-\frac{b}{\left\|\hat{p}_{ij}\right\|^4} - \frac{a}{4D^2 - \left\|\hat{p}_{ij}\right\|^2}\right) \tag{6.40}$$

Specially, it should be noted that parameters a and b at least satisfy the constraint condition $\frac{\sqrt{b^2+16abD^2}-b}{2a} < D^2$. Besides, the formation flocking task requires that the potential function $V\left(\left\|\tilde{p}_{ij}(t)\right\|\right)$ reaches its minimal value at the point of $\left\|\tilde{p}_{ij}(t)\right\| = \frac{\left\|\hat{p}_{ij}(t)\right\|}{\left\|p_{ij}^d\right\|} = 1$. Thus, the parameter \$a\$ and \$b\$ should also satisfy another constraint condition $a = b\left(4D^2 - 1\right)$. Figure 6.3 gives an example of the potential function, which satisfies the three rules of Definition 5.1. Agents with generic initial conditions are employed, on condition that the initial interaction network among the followers is an undirected connected graph and there exists at least one follower having a leader neighbor at the initial time. According to Theorem 6.1 and Corollary 6.1, the connectivity of the interaction network can be preserved at all times.

Firstly, we assign five agents to achieve the leader–follower cohesive flocking task. Leader swims with bounded varying speeds that$f_1(t) = 0.1/(t+1)$ and$f_2(t) = 0.1/(t+1)$, while followers try to track the velocities of the leader and construct a cohesive formation together with the leader at the same time. The simulation results are shown in Fig. 6.4. In Fig. 6.4a, the green star point is the initial position of each agent, while the centre of the colored ball denotes the final position of each agent at time $t = 200$ s, which has been enlarged for more details. The rough red circle line denotes the trajectory of the leader. It is clear to see that the five agents asymptotically

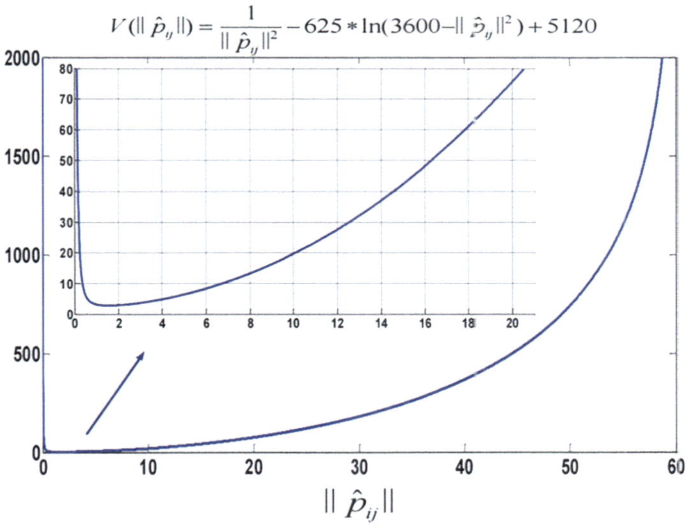

Fig. 6.3 Potential function (6.39), $a = 625$, $b = 1$, $c = 5120$, $D = 30$

converge to a stable flocking formation. Figure 6.4b, c tell us that the thrusting speed, the rotational speed, and the heading angles of the followers asymptotically track those of the leader. Figure 6.4d gives the distances between any two agents. The distances are all asymptotically stable. Besides, the initial adjacency matrix of the system is

$$A(0) = \begin{bmatrix} 1&0&0&0&0&0&0&0&0&0 \\ 1&0&1&1&1&0&1&1&1&1 \\ 1&1&0&1&1&1&1&1&1&1 \\ 1&1&1&0&1&1&1&1&1&1 \\ 1&1&1&1&0&1&1&1&1&1 \\ 0&0&1&1&1&0&1&1&1&1 \\ 1&1&1&1&1&1&0&1&1&1 \\ 0&1&1&1&1&1&1&0&1&1 \\ 1&1&1&1&1&1&1&1&0&1 \\ 1&1&1&1&1&1&1&1&1&0 \end{bmatrix} \tag{6.41}$$

which indicates that the followers' interaction network is a complete graph, and only part of the followers have a leader neighbor. We output the adjacency matrix of the whole system in real time, according to which we conclude that the connectivity of the system is preserved at all times. Finally, the interaction network of the whole system is fixed, which can be expressed by the following adjacency matrix

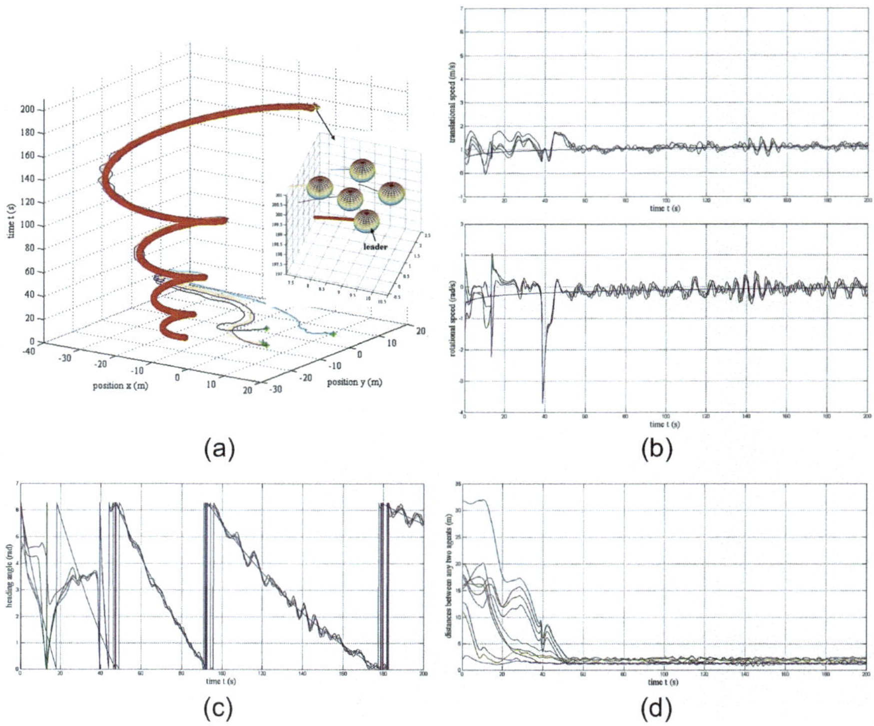

Fig. 6.4 Simulation results of the leader–follower cohesive flocking task. **a** Trajectories of the five agents. **b** The thrusting speeds and the rotational speeds of the five agents with time t. **c** The heading angles of the five agents with time t. **d** The distances between any two agents with time t

$$A(200) = \begin{bmatrix} 1 & 0 & 0 & 0 & 0 & 0 & 0 & 0 & 0 & 0 \\ 1 & 0 & 1 & 1 & 1 & 1 & 1 & 1 & 1 & 1 \\ 1 & 1 & 0 & 1 & 1 & 1 & 1 & 1 & 1 & 1 \\ 1 & 1 & 1 & 0 & 1 & 1 & 1 & 1 & 1 & 1 \\ 1 & 1 & 1 & 1 & 0 & 1 & 1 & 1 & 1 & 1 \\ 1 & 1 & 1 & 1 & 1 & 0 & 1 & 1 & 1 & 1 \\ 1 & 1 & 1 & 1 & 1 & 1 & 0 & 1 & 1 & 1 \\ 1 & 1 & 1 & 1 & 1 & 1 & 1 & 0 & 1 & 1 \\ 1 & 1 & 1 & 1 & 1 & 1 & 1 & 1 & 0 & 1 \\ 1 & 1 & 1 & 1 & 1 & 1 & 1 & 1 & 1 & 0 \end{bmatrix} \tag{6.42}$$

Secondly, we verify the leader–follower formation flocking algorithm by five gents similarly. For the sake of simplicity, we only show the trajectory of the five agents to complete the leader–follower ring-shaped flocking task in Fig. 6.5, and the final position of the five agents has also been enlarged for more details.

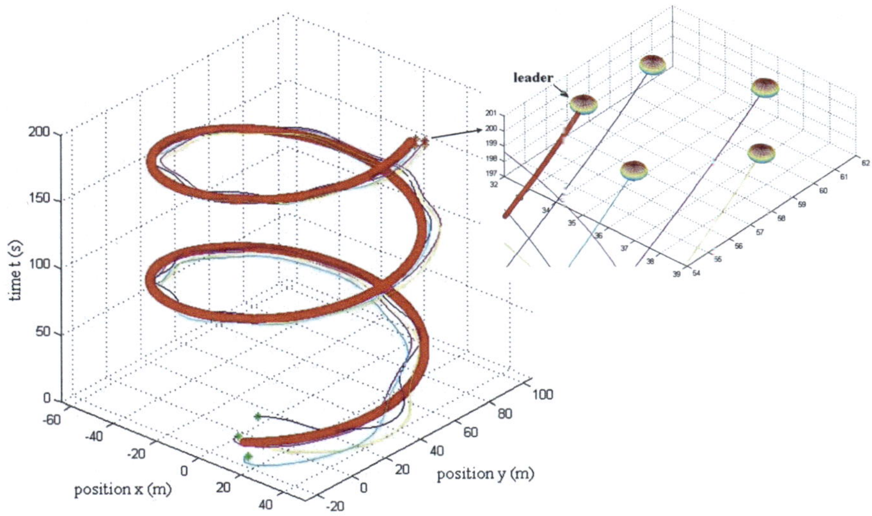

Fig. 6.5 Simulation results of the leader–follower ring-shaped formation flocking task

6.3 Summary

A distributed leader–follower adaptive flocking problem of multiple robotic fish governed by extended second-order unicycles has been investigated in this chapter. Consider the system consisting of only one leader with varying but bounded speeds. Two leader–follower adaptive flocking algorithms are proposed with the combination of consensus and attraction/repulsion functions to respectively solve the cohesive flocking problem and the formation flocking problem. Provided that the initial interaction network among the followers is an undirected connected graph, and there exists at least one follower having a leader neighbor at the initial time, the followers asymptotically converge to a cohesive flocking configuration or formation flocking configuration together with a variable-speed leader. The stability of the closed-loop system is analyzed by LaSalle-Krasovskii invariance principle. Finally, simulation results verify the effectiveness of the proposed control algorithms.

References

1. Gu, D., & Wang, Z. (2009). Leader-follower flocking: Algorithms and experiments. *IEEE Transactions on Control Systems Technology, 17*, 1211–1219.
2. Yu, W., Chen, G., & Cao, M. (2010). Distributed leader-follower flocking control for multi-agent dynamical systems with time-varying velocities. *Systems and Control Letters, 59*. 543–552.

3. Savkin, A. V., & Teimoori, H. (2009). Decentralized formation flocking and stabilization for networks of unicycles. In *Proceedings of the 48th IEEE Conference on Decision Control and the 28th Chinese Control Conference* (pp. 984–989).
4. Hou, Z., Cheng, L., & Tan, M. (2009). Decentralized robust adaptive control for the multi-agent system consensus problem using neural networks. *IEEE Transactions on Systems, Man, and Cybernetics. Part B, Cybernetics, 39*, 636–647.
5. Demetriou, M. A. (2010). Design of consensus and adaptive consensus filters for distributed parameter systems. *Automatica, 46*, 300–311.
6. Min, H., Sun, F., Wang, S., & Li, H. (2011). Distributed adaptive consensus algorithm for networked Euler-Lagrange systems. *IET Control Theory and Applications, 5*, 145–154.
7. Su, H., Chen, G., Wang, X., & Lin, Z. (2011). Adaptive second-order consensus of networked mobile agents with nonlinear dynamics. *Automatica, 47*, 368–375.
8. Liu, Y., & Jia, Y. (2012). Adaptive consensus protocol for networks of multiple agents with nonlinear dynamics using neural networks. *Asian Journal of Control, 14*, 1328–1339.
9. Latombe, J. C. (1991). *Robot motion planning*. Kluwer Academic Publishers.
10. Zhang, D., Wang, L., Yu, J., & Tan, M. (2007). Coordinated transport by multiple biomimetic robotic fish in underwater environment. *IEEE Transactions on Control Systems Technology, 15*, 658–671.
11. Shao, J., Xie, G., & Wang, L. (2007). Leader-following formation control of multiple mobile vehicles. *IET Control Theory and Applications, 1*, 545–552.
12. Klein, D. J., Bettale, P. K., Triplett, B. I., & Morgansen, K. A. (2008). Autonomous underwater multivehicle control with limited communication: Theory and experiment. In *International Federation of Automatic Control Workshop on Navigation, Guidance and Control of Underwater Vehicles*.
13. Tanner, H. G. (2004). On the controllability of nearest neighbour interconnections. In *43rd IEEE Conference on Decision Control* (Vol. 3, p. 2467).
14. Su, H., Wang, X., & Chen, G. (2010). Rendezvous of multiple mobile agents with preserved network connectivity. *Systems and Control Letters, 59*, 313–322.
15. Zavlanos, M. M., Jadbabaie, A., & Pappas, G. J. (2007). Flocking while preserving network connectivity. In *46th IEEE Conference on Decision Control* (pp. 2919–2924).
16. Khalil, H. K. (2002). *Nonlinear systems* (third ed.). Prentice Hall.

Chapter 7
Application Scenarios and Countermeasures of Distributed Coordination Control System Considering Failure Individuals

7.1 Typical Application Scenarios

In 2015, Intel combined curiosity and creativity to explore the potential of drones. Intel's CEO Brian Krzanich, along with Anil Nanduri, head of the perception computing market, a group of artists, and researchers from Ars Electronica Futurelab, created an outdoor drone light show with live music in Hamburg, Germany. In November 2015, 100 drones equipped with lights danced across the night sky in Hamburg, drawing beautiful 3D patterns and writing elegant text (see Fig. 7.1). Nanduri believe that a new era for drones had arrived.

Horst Hoertner, Senior Director at Intel Futurelab, stated that to turn Intel's vision into reality, they focused on drone swarm technology. Hoertner leads a 15-person team that refers to these drones as "spaxels," meaning "space pixels." Before the drones took flight, engineers used ground control software to design the flight paths, final light show patterns, and light effects synchronized with the symphony. Hoertner expressed a desire to use drones to create aesthetic and socially meaningful experiences. Two months later, Krzanich premiered a video at CES 2016, showcasing 100 drones flying in a new and safe way to reinvent the fireworks experience.

The drone light show utilizes swarm control technology to reshape artistic beauty and visual experience. The emergence of light shows has demonstrated the vast potential of swarm control technology. Since then, several prominent domestic companies, including Shenzhen Zero Degree, Zero Degree Intelligence, JD Logistics, ZhenDi Technology, EHang, and DaMo DaZhi, have entered the drone light show market, leveraging mature drone technology and extensive market potential.

On April 29, 2018, Guangdong EHang Baiju Media Technology Co., Ltd. (hereafter referred to as "EHang Baiju") launched 1374 drones simultaneously, illuminating the ancient city wall of Xi'an. The 1374 drones symbolized the shared development of China with 74 countries and international organizations under the "Belt and Road" initiative proposed in 2013; represented the historical city wall of Xi'an, which has a circumference of 13.74 km and is the oldest and best-preserved ancient wall

Y. Jia, *Distributed Coordination Control System*,
https://doi.org/10.1007/978-981-96-0365-7_7

Fig. 7.1. Intel light show 2015

structure; and commemorated the large-scale expansion and construction that began in 1374. The performance set a Guinness World Record for the most drones flying simultaneously. However, during the "Red May · City Wall International Cultural Festival" debut on May 1, the drone light show experienced errors, with some patterns displaying "garbled" images. EHang Baiju's investigation found that while the drone equipment, communication systems, and flight systems were functioning normally, some drones experienced directional interference with their positioning and auxiliary positioning systems after takeoff, leading to data anomalies in their positions and altitudes. This light show not only showcased the immense potential of future swarm control technology but also highlighted the current technological immaturity. Most current light shows still use a centralized control system, relying on ground stations for task planning and inter-drone communication, lacking inter-drone information exchange. Consequently, directional interference can disrupt planned operations. Future solutions will need mature distributed control technology to address this issue effectively.

7.2 Countermeasure Research

The solutions discussed in this chapter are aimed at multi-agent systems with the following characteristics: each agent has equal communication capabilities, equal positioning capabilities, and a consistent decision-making mechanism.

The main goal is to address situations where some agents in a swarm experience faults during collaborative tasks. Specifically, when some agents fail, the remaining agents must be able to reconfigure into a new cooperative formation and continue performing the task. This solution is particularly applicable to scenarios where unmanned systems encounter sudden failures during search, rescue, or exploration missions.

This chapter presents a specific solution using the example of drone escort missions.

7.2.1 Introduction

Aerial escort task faces to multiple aerial vehicles that are desired to follow certain aerial vehicle (named target aerial vehicle) or protect this target aerial vehicle from being detected. Recent years, the rapid development of unmanned aerial vehicles (UAVs) and related technologies brings new opportunities and potential application directions to escort task. Miniature UAVs owning these advantages of low-cost and low-risk, become a better solution in implementing escort task as compared with manned aerial vehicles [1, 2].

Under this background, an escort system consists of at least one target aerial vehicle and several miniature UAVs. The target aerial vehicle usually refers to the one equipped with sophisticated devices, and almost takes the domination power in grasping the global situation. These miniature UAVs are equipped with necessary devices only, and are introduced for the purpose of protecting the target aerial vehicle.

In order to protect the target aerial vehicle from being detected, two properties should be considered. First, these miniature UAVs need to act as a complicated and changeable entirety aligning with the target aerial vehicle and thus provide a time-variant relative position of the target aerial vehicle to the whole system. The accurate position of the target aerial vehicle will be time-variant and be not easy to be detected. Secondly, the target aerial vehicle is surrounded by these miniature UAVs, which are taken as the spatial expansion of the target aerial vehicle. Then, it becomes not so easy to reach the location of the target aerial vehicle. In a word, a distributed coordination control method is required for these miniature UAVs to realize the proposed two properties when executing the aerial escort task.

During the aerial escort process, the target aerial vehicle autonomously makes decisions according to the surrounding environment, and the trajectory of the target aerial vehicle has no concern with these miniature UAVs. For this reason, the target aerial vehicle is considered as a leader. As the protected object, the target aerial vehicle's position information and attitude angle information should be received by at least one miniature UAV through the communication network built by the nearest neighbor interaction rule [3]. These miniature UAVs are considered as the followers. Followers are expected to build a desired formation bench-marked against the position of the target aerial vehicle while avoid collisions with each other. To this end, the aerial escort task involved in this chapter can be simplified as a three-dimensional formation control problem of multi-agent system with the leader-follower structure.

Formation control of multi-agent system is a perennial problem that has been intensively studied. Except for these traditional methods including leader-follower [4, 5], behavior-based [6–8], and virtual structure [9–11], new algorithms are also investigated, such as adaptive method [12–14], neural network control [15], sliding mode control [16], model predictive control [17], and distributed control [18]. Furthermore, the combination of these methods is also proposed to solve specific formation control problems. Therein, the leader-follower structure is considered to be an effective method for engineering applications and be easy to combine with other algorithms [19, 20]. For example, the leader-follower structure and an adaptive method are combined to solve the formation control problem based on vision [21, 22]. Leader-follower structure and flocking algorithm are combined to solve the two-dimensional formation control problem of large-scale agents [23].

Flocking is a kind of collective phenomena, which can be widely observed from flocks of birds, schools of fish, herds of quadrupeds, and swarms of bees. Because of its autonomy, self-organization, scalability, flexibility, and robustness [24–27], flocking algorithm is a suitable option for cohesive escort tasks of miniature UAVs. Flocking algorithm also can be used to realize desired formation control of multi-agent system [23], such as sphere configuration and herringbone configuration. The sphere escorting task is the smallest configuration that offers the target aerial vehicle with an omnidirectional protection, while herringbone aerial escort task is exactly a typical application of the three-dimensional migrating flocking behaviors of wild gooses.

In this paper, the target aerial vehicle and the miniature UAVs are all simplified as the same dynamic model with limited communication capability. Graph theory is used to depict the escort system's communication network [28], based on which a flocking protocol is proposed for miniature UAVs. These miniature UAVs are forced to fly under the guidance of the target aerial vehicle and to form an expected geometrical configuration together with the target aerial vehicle. Moreover, two safe problems involved in aerial escort task are also discussed. First, the robustness of the system is discussed in case part of these miniature UAVs are invalid. Secondly, during the escort process, aerial vehicles are capable of avoiding collision with each other. Finally, three classes of aerial escort tasks, including the cohesive escort task, the sphere escort task, and the herringbone escort task, are illustrated by one target aerial vehicle and several miniature UAVs. Numerical simulations are carried out to validate the functionalities of the proposed algorithm for these specific aerial escort tasks.

7.2.2 Modeling

The escort system consists of one target aerial vehicle and $N-1$ miniature UAVs. Therein, $\mathbb{F} = \{1\}$ denotes the leader set including only one target aerial vehicle, and $\mathbb{F} = \{2, \ldots, N\}$ denotes the follower set including $N-1$ miniature UAVs. The minimum turning radius of each aerial vehicle equals to R. Then, collision will not happen between any two individuals, if the distance is larger than $2R$. $2R$ is called as safe distance.

Let $\theta_i(t)$ be the yaw angle, $\phi_i(t)$ be the pitch angle, $v_i(t)$ be the thrusting speed, $\omega_i(t)$ be the rotational speed, $l_i(t) > 0$ be a positive constant that denotes the distance between the geometrical center and the mass center, $a_i(t)$ be the thrusting acceleration, $b_i(t)$ be the rotational acceleration, and $c_i(t)$ be the pitch angle speed, $i = 1, \ldots, N$. Besides, unit vectors

$$\overrightarrow{t_i}(t) = [\cos\phi_i(t)\cos\theta_i(t), \sin\phi_i(t), \cos\phi_i(t)\sin\theta_i(t)]^T \tag{7.1}$$

and

$$\overrightarrow{n_i}(t) = [-\cos\phi_i(t)\sin\theta_i(t), 0, \cos\phi_i(t)\cos\theta_i(t)]^T \tag{7.2}$$

are orthogonal to each other. Without loss of generality, the target aerial vehicle and the miniature UAVs are all modeled as the following kinematic equation

$$\begin{aligned} \dot{p}_i(t) &= H_i^T(t)q_i(t) \\ \dot{q}_i(t) &= u_i(t) \\ \dot{\phi}_i(t) &= c_i(t) \end{aligned} \tag{7.3}$$

where position vector $p_i(t) = \left[x_i(t), y_i(t), z_i(t)\right]^T$, $H_i(t) = \left[\overrightarrow{t_i}, \overrightarrow{n_i}\right]^T$, velocity vector $q_i(t) = [v_i(t), \omega_i l_i(t)]^T$, and $u_i(t) = [a_i(t), b_i(t)]^T$. The control input of each individual is denoted by $uu_i(t) = [a_i(t), b_i(t), c_i(t)]^T, i = 1, ..., N$.

7.2.3 Communication Mechanism Design

Nearest neighbor interaction rule is adopted for the escort system to build the communication network $G(t)$ by a directed graph ξ with node set $\{1, 2, \ldots, \mathrm{N}\}$ and a time-varying edge set $\varepsilon \in v \times v$. ε denotes the neighbor relationships between any two aerial vehicles. There are two basic rules. For a miniature UAV, interconnection with the target aerial vehicle is unidirectional. The interconnection between any two miniature UAVs is bidirectional. Therefore, the information of the target aerial vehicle can be obtained by the miniature UAV if the miniature UAV has a leader neighbor, but the behavior of the miniature UAV won't be adopted by the target aerial vehicle.

The neighbor set $N_i(t)$ of miniature UAV i at initial time $t = 0$ is denoted by

$$N_i(0) = \left\{j \mid ||P_i(0) - P_j(0)|| < D, i, j = 1, \ldots, N, j \neq i\right\} \tag{7.4}$$

where $||\cdot||$ is the Euclidean norm, and $D > 2R$ is the communication radius.

In order to clarify the neighbor relationship, the adjacency matrix $A_N(t) = \left[\omega_{ij}(t)\right]_{N\times N}$ of the graph $G(t)$ is introduced as follows:

$$w_{ij}(t) = \begin{cases} 1, \ i = 1, j = 1 \\ 1, \ i \in \mathbb{F}, j \in N_i(t) \\ 0, \ otherwise \end{cases} \tag{7.5}$$

The communication network $G(t)$ is supposed to switch at t_p, $p = 1, 2\ldots$ One property of the network should be seriously considered is connectivity. Then, in order to keep the connectivity of the network, one hysteresis is introduced [23]:

(1) if $(i,j) \in \varepsilon(t^-)$ and $\|\hat{p}_i(t) - \hat{p}_j(t)\| < 2D$, where $\hat{p}_i(t) = p_i(t) - p_i^*(t)$ and $p_i^*(t) = \int_0^t w_{i1} H_i^T(t) q_1 d\tau (i = 1, \ldots, N)$, then $(i,j) \in \varepsilon(t)$, for $t > 0$;
(2) if $(i,j) \notin \varepsilon(t^-)$ and $\|p_i(t) - p_j(t)\| < D$, then $(i,j) \in \varepsilon(t)$, for $t > 0$.

7.2.4 Distributed Coordination Control Algorithm Design and Stability Analysis

Since the main problem is how to design the coordination control protocol for each miniature UAVs under the guidance of the target aerial vehicle, the trajectory of the target aerial vehicle can be simplified as a spatial one with zero-value control input. That is,

$$uu_1(t) = \begin{bmatrix} a_1(t) \\ b_1(t) \\ c_1(t) \end{bmatrix} = 0 \tag{7.6}$$

Here, **0** denotes the zero vector. It should be noted that initial rotational speed $\omega_1 \neq 0$ and initial pitch angle $\phi_1 \neq 0$. Besides, in order to save space, we may use p_i instead of $p_i(t)$ for example in the following sections.

Let $\widehat{\theta_i} = \theta_i - \omega_{i1}\theta_1$, $\widehat{\phi_i} = \phi_i - \omega_{i1}\phi_1$, and $\widehat{q_i} = q_i - \omega_{i1}q_1$. Coupled with (7.3), one gets

$$\dot{\hat{p}}_i = \dot{p}_i - \dot{p}_i^* = H_i^T q_i - w_{i1} H_i^T q_1 = H_i^T q_1 = H_i^T \hat{q}_i, \quad i = 1, \ldots, N. \tag{7.7}$$

Then, the cooperative protocol of miniature UAV i for three-dimensional cohesive aerial escort task is designed by

$$\begin{aligned} a_i &= -\vec{t}_i^T \sum_{j \in N_i} \left(\dot{\hat{p}}_i - \dot{\hat{p}}_j \right) - \vec{t}_i^T \sum_{j \in N_i} \nabla_{\hat{p}_{ij}} V\left(\|\hat{p}_{ij}\|\right) \\ b_i &= -l_i \sum_{j \in N_i} \left(\hat{\theta}_i - \hat{\theta}_j\right) - \vec{n}_i^T \sum_{j \in N_i} \left(\dot{\hat{p}}_i - \dot{\hat{p}}_j \right) - \vec{n}_i^T \sum_{j \in N_i} \nabla_{\hat{p}_{ij}} V\left(\|\hat{p}_{ij}\|\right) \\ c_i &= -\sum_{j \in N_i} \left(\hat{\phi}_i - \hat{\phi}_j\right) \end{aligned} \tag{7.8}$$

where i $\in \mathbb{F}$ and $\hat{p}_{ij} = \hat{p}_i - \hat{p}_j$. $\nabla_{\hat{p}_{ij}} V\left(\|\hat{p}_{ij}\|\right)$ is the gradient of an artificial potential function $V\left(\|\hat{p}_{ij}\|\right)$, which is a differentiable, nonnegative, radially unbounded function of the distance $\|\hat{p}_{ij}\|$ between agent i and j, such that

(1) $V\left(\left\|\hat{p}_{ij}\right\|\right) \to \infty$ as $\left\|\hat{p}_{ij}\right\| \to 2R^*$;
(2) $V\left(\left\|\hat{p}_{ij}\right\|\right) \to \infty$ as $\left\|\hat{p}_{ij}\right\| \to 2D$;
(3) $V\left(\|\hat{p}_{ij}\|\right)$ attains its unique minimum when $\|\hat{p}_{ij}\|$ equals to a desired value between $2R^*$ and 2D.

The total potential is defined as

$$V = \sum_{i \in \mathbb{F}} \sum_{j \in N_i} V\left(\|\hat{p}_{ij}\|\right) + \sum_{j \in N_l} V\left(\|\hat{p}_{1j}\|\right) \tag{7.9}$$

where $N_l = \{i | w_{i1} = 1, i \in \mathbb{F}\}$ denotes the set of miniature UAVs that have one target aerial vehicle neighbor.

To this end, the basic results are stated on the cohesive escort task of the escort system flying in three-dimensional space as follows.

Theorem 7.1 (Three-dimensional Cohesive Aerial Escort) *Consider an escort system consisting of one target aerial vehicle and $N - 1$ miniature UAVs with kinematics (7.3). The target aerial vehicle and miniature UAVs are controlled by protocols (7.6) and (7.8). Suppose that the initial communication network $G(0)$ is leader-follower connected. Then the following conclusions hold*:

1. *The connectivity of the communication topology is preserved at all times.*
2. *No collision happens between neighbors.*
3. *The speed and the attitude angle of each miniature UAV asymptotically become the same as those of the target aerial vehicle.*
4. *The escort system approaches a cohesive formation when the total potential is minimized.*

Given a geometric pattern χ that owns N vertices $p_i^d = \left[x_i^d, y_i^d\right]^T$, $i = 1, ..., N$. Then, the control protocol of miniature UAV $i (i \in \mathbb{F})$ for formation aerial escort task is rewritten as

$$\begin{aligned} a_i &= -\vec{t}_i \sum_{j \in N_i} \left(\dot{\hat{p}}_i - \dot{\hat{p}}_j\right) - \vec{t}_i \sum_{j \in N_i} \nabla_{\tilde{p}_{ij}} V\left(\tilde{p}_{ij}\right) \\ b_i &= -l_i \sum_{j \in N_i} \left(\hat{\theta}_i - \hat{\theta}_j\right) - \vec{n}_i \sum_{j \in N_i} \left(\dot{\hat{p}}_i - \dot{\hat{p}}_j\right) - \vec{n}_i \sum_{j \in N_i} \nabla_{\tilde{p}_{ij}} V\left(\tilde{p}_{ij}\right) \\ c_i &= -\sum_{j \in N_i} \left(\hat{\phi}_i - \hat{\phi}_j\right) \end{aligned} \tag{7.10}$$

where $\tilde{p}_{ij} = \frac{\left\|\widehat{p_{ij}}\right\|}{\left\|p_{ij}^d\right\|}$ and $p_{ij}^d = p_i^d - p_j^d$.

Corollary 7.1 (Three-dimensional Formation Aerial Escort) *Consider an escort system consisting of one target aerial vehicle and $N - 1$ miniature UAVs with kinematics (7.3). The target aerial vehicle and miniature UAVs are controlled by protocols (7.6) and (7.10). Suppose that the initial communication network $G(0)$ is leader-follower connected. Then the following conclusions hold:*

1. *The connectivity of the communication topology is preserved at all times.*

2. *No collision happens between neighbors.*
3. *The speed and the attitude angle of each miniature UAV asymptotically become the same as those of the target aerial vehicle.*
4. *The escort system approaches a geometric formation χ when the total potential is minimized.*

Both the proof processes of Theorem 7.1 and Corollary 7.1 are omitted here due to the limited space. Readers can look through another paper [23], in which a similar proof process has been given.

7.2.5 Simulation Experimental Validation and Discussion

7.2.5.1 Cohesive Aerial Escort Task

Firstly, the effectiveness of the proposed control protocols (7.6) and (7.8) for one target aerial vehicle and five miniature UAVs to execute the cohesive aerial escort task in three-dimensional space is going to be analyzed. Secondly, take cohesive aerial escort task for example, the robustness of the escort system needs to be discussed in further. Suppose that No.2 miniature UAV drops from the group because of internal or external factors, such as the equipment malfunction from itself or the gunfire from the enemy. And then, whether the remaining four miniature UAVs will intelligently adjust their behaviors and continue to complete the cohesive escort task or not? In brief, the numerical validation process should be divided into two stages, before No.2 miniature UAV drops and after No.2 miniature UAV drops. The first stage aims to verify the effectiveness of the flocking algorithm (7.8), while the second stage focuses on discussing the robustness of the escort system.

Consider one target aerial vehicle and five miniature UAVs respectively controlled by the protocols (7.6) and (7.8) with the following initial states

$$
\begin{aligned}
& p_1(0) = [-5.7\,\text{m}, -2.4\,\text{m}, -2.1\,\text{m}]^T, \theta_1(0) = -0.35\text{ rad}, \\
& \emptyset_1(0) = -0.91\text{ rad}, v_1(0) = 1.2\text{ m/s}, \omega_1(0) = 0.15\text{ rad/s} \\
& p_2(0) = [-17.6\,\text{m}, 1.3\,\text{m}, 10.5\,\text{m}]^T, \theta_2(0) = 2.8\text{ rad}, \\
& \emptyset_2(0) = 0.11\text{ rad}, v_2(0) = 0.6\,\text{m/s}, \omega_2(0) = -0.25\text{ rad/s} \\
& p_3(0) = [-1\,\text{m}, -6.7\,\text{m}, -11\,\text{m}]^T, \theta_3(0) = -0.34\text{ rad}, \\
& \emptyset_3(0) = -0.84\text{ rad}, v_3(0) = 0.27\,\text{m/s}, \omega_3(0) = -0.01\text{ rad/s} \\
& p_4(0) = [-2.6\,\text{m}, -12.3\,\text{m}, 6.6\,\text{m}]^T, \theta_4(0) = 0.17\text{ rad}, \\
& \emptyset_4(0) = 0.27\text{ rad}, v_3(0) = 0.43\,\text{m/s}, \omega_4(0) = -0.09\,\text{rad/s} \\
& p_5(0) = [-1\,\text{m}, 18.3\,\text{m}, -18.7\,\text{m}]^T, \theta_5(0) = 1.28\text{ rad}, \\
& \emptyset_5(0) = 0.18\text{ rad}, v_3(0) = 1.79\,\text{m/s}, \omega_5(0) = -0.1\text{ rad/s} \\
& p_6(0) = [-9\,\text{m}, -10.8\,\text{m}, -7.3\,\text{m}]^T, \theta_6(0) = 0.76\text{ rad},
\end{aligned}
$$

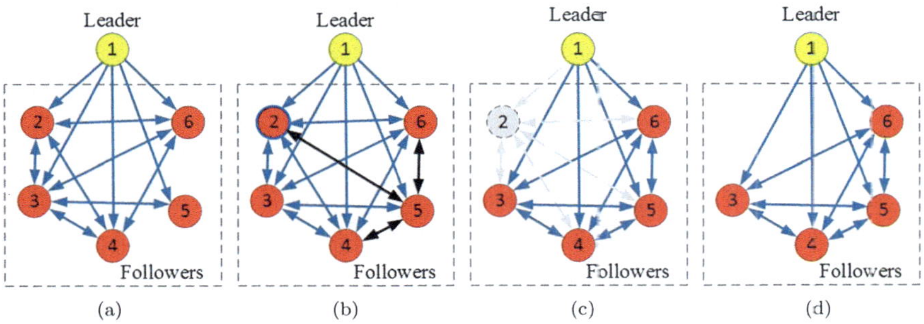

Fig. 7.2. Topologies of one target aerial vehicle and five to four miniature UAVs during cohesive aerial escort task. **a** $t = 0$ s, **b** $t = 500^{-}$ s, **c** $t = 500^{+}$ s, $(d) t = 1000$ s

$$\varnothing_6(0) = 0.45\ \text{rad},\ v_6(0) = 0.14\ \text{m/s},\ \omega_6(0) = -0.14\ \text{rad/s} \tag{7.11}$$

The communication radius D = 30 m.

According to the communication mechanism and the initial conditions of these aerial vehicles, the initial topology of the escort system is shown in Fig. 7.2a. The yellow node denotes the target aerial vehicle (that is, the leader), while the red ones denote the miniature UAVs (that is, the followers). The initial topology should be a leader-follower connected graph, but not necessary a complete one.

Figure 7.3 gives the trajectories of the aerial vehicles during the evolution process of cohesive escort task under the control protocols (7.6) and (7.8). The green star points denote their initial positions $p_1(0), \ldots, p_6(0)$. The bold red line denotes the trajectory of the target aerial vehicle. The bigger color ball denotes the target aerial vehicle, and the smaller color balls denote the miniature UAVs.

(1) Cohesive Escort Task

Figure 7.2b shows that the communication topology of the escort system already become a complete one at time $t = 500^{-}$ s. The small picture at the right corner of Fig. 7.3 shows the positions of the target aerial vehicle and five miniature UAVs at time $t = 500^{-}$ s, which proves that six aerial vehicles have formed a desired cohesive configuration. Figure. 7.4 gives more details about the distances between any two aerial vehicles from $t = 0$ s to $t = 500^{-}$ s. It is obvious that one target aerial vehicle and five miniature UAVs converge to a steady configuration around $t = 100$ s, and the steady state of the system is maintained almost from $t = 100$ s to $t = 500^{-}$ s. Figure 7.5 respectively give the variation of the yaw angle, the pitch angle, the thrusting speed, and the rotational speed versus time during the cohesive escort task. It is clear to see that the attitude angle and the speed of each miniature UAV converges to the same with that of the target aerial vehicle before $t = 500$ s.

(2) Robustness of the Escort system

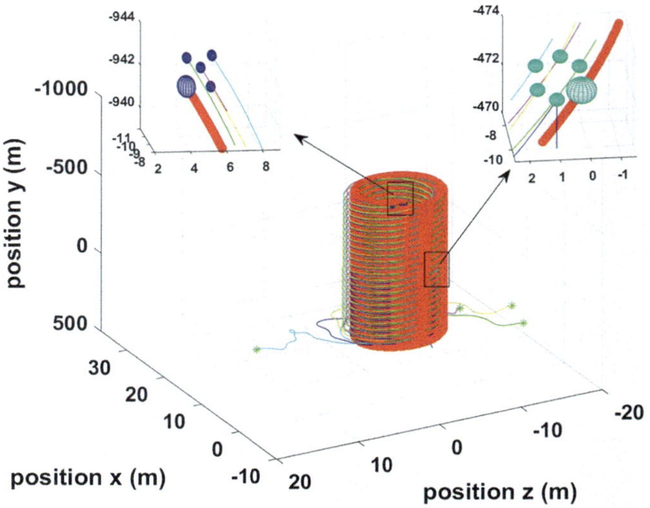

Fig. 7.3. Simulation results of the trajectories of one target aerial vehicle and from five to four miniature UAVs.

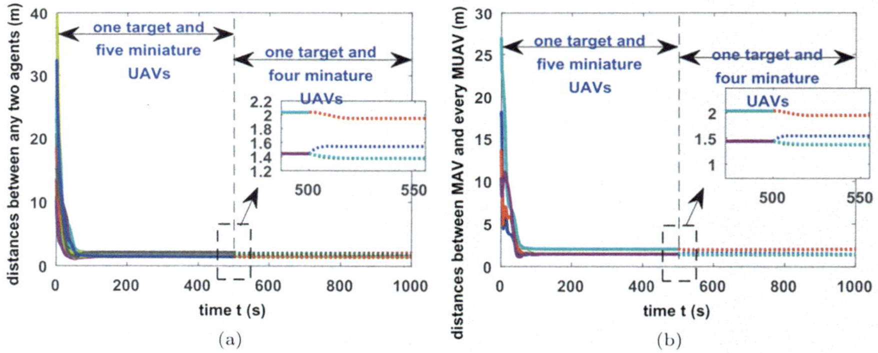

Fig. 7.4. Simulation results of the distances of one target aerial vehicle and from five to four miniature UAVs. **a** Distances between any two agents. **b** Distances between target aerial vehicle and every miniature UAV

As mentioned above, suppose that No.2 miniature UAV is invalid at $t = 500$ s. Then, the steady state of the escort system is thus destroyed and the configuration of the escort system begins to be rebuilt under the effort of protocol (7.8).

At the same time, the communication topology of the air system is also changed at time $t = 500$ s. All the edges relative to node 2 are all vanished as shown in Fig. 7.2c. The final communication topology of the escort system become another complete graph as shown in Fig. 7.2d. The left small picture in Fig. 7.3 gives the

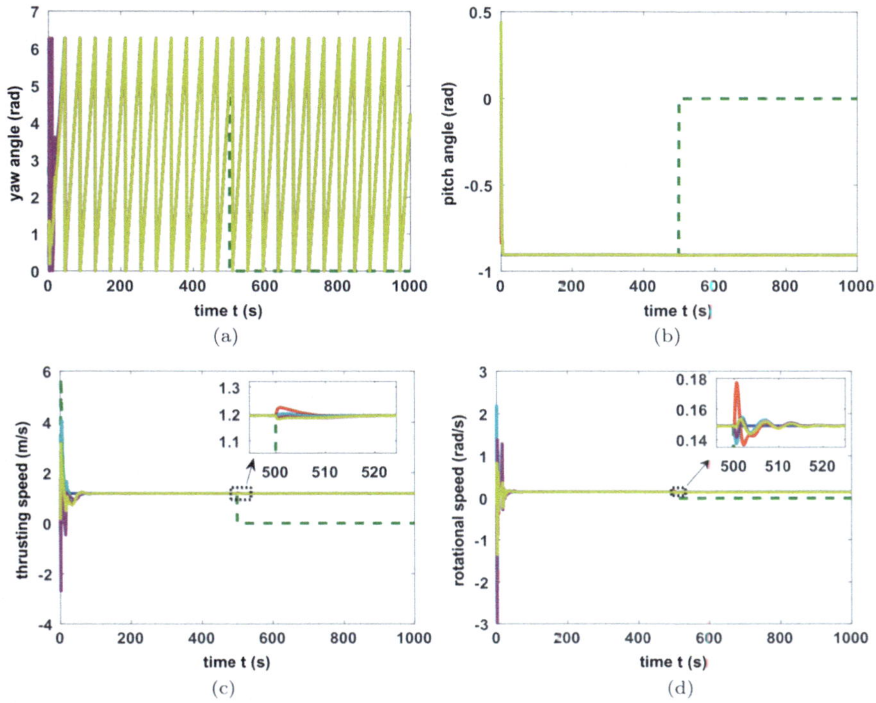

Fig. 7.5 Simulation results of the attitude angle and the speed of one target aerial vehicle and from five to four miniature UAVs. **a** Yaw angle versus time. **b** Pitch angle versus time. **c** Thrusting speed versus time. **d** Rotational speed versus time

final configuration of the new system at time $t = 1000$ s, in which the remaining four miniature UAVs form a new cohesive configuration around the target aerial vehicle.

When No.2 miniature UAV is invalid at $t = 500$ s, all the attitude angles and the speeds of No.2 miniature UAV become zero as shown in Fig. 7.5, which are expressed by the green dashed lines. Then, the attitude angles and the speeds of the remaining four miniature UAVs need to be adjusted for the purpose of reaching a new configuration of the escort system. Figures 7.4 and 7.5c, d give more details on the building of new balance state. Under the effort of the control protocols (7.6) and (7.8), the escort system converges to another steady state finally. In addition, Fig. 7.4 also verifies that no collision happens during the whole evolution process.

7.2.5.2 Formation Aerial Escort Task

In this section, two classical formation aerial escort tasks including sphere escort and herringbone escort are discussed in order to test the effectiveness of the control protocols (7.6) and (7.10). Sphere aerial escort task requires the miniature UAVs to fly around the target aerial vehicle for the purpose of protecting the target aerial vehicle from being attacked. Herringbone aerial escort task is inspired from the formation flying of the wild gooses, and the miniature UAVs are used as decoys to puzzle enemy with a wrong central position of the target aerial vehicle.

(1) **Sphere Aerial Escort Task**

Sphere aerial escort is definitely a formation control problem that forces the miniature UAVs to locate dispersedly at the surface of a ball whose center is the position of the target aerial vehicle. Therefore, we suppose that the desired formation information is

$$
\begin{aligned}
&p_1^d = r.[0, 0, 0]^T, p_2^d = r.[0, 0, 1]^T, p_3^d = r.[0, 0, -1]^T\\
&p_4^d = r.\left[\cos\left(\frac{\pi}{4}\right)\cos\left(\frac{2\pi}{3}\right), \cos\left(\frac{\pi}{4}\right)\sin\left(\frac{2\pi}{3}\right), \sin\left(\frac{\pi}{4}\right)\right]^T\\
&p_5^d = r.\left[\cos\left(\frac{\pi}{4}\right)\cos\left(\frac{4\pi}{3}\right), \cos\left(\frac{\pi}{3}\right)\sin\left(\frac{4\pi}{3}\right), \sin\left(\frac{\pi}{4}\right)\right]^T\\
&p_6^d = r.\left[\cos\left(\frac{\pi}{4}\right)\cos(2\pi), \cos\left(\frac{\pi}{4}\right)\sin(2\pi), \sin\left(\frac{\pi}{4}\right)\right]^T\\
&p_7^d = r.[\cos(0)\cos(0), \cos(0)\sin(0), \sin(0)]^T\\
&p_8^d = r.\left[\cos(0)\cos\left(\frac{\pi}{2}\right), \cos(0)\sin\left(\frac{\pi}{2}\right), \sin(0)\right]^T\\
&p_9^d = r.[\cos(0)\cos(\pi), \cos(0)\sin(\pi), \sin(0)]^T\\
&p_{10}^d = r.\left[\cos(0)\cos\left(\frac{3\pi}{2}\right), \cos(0)\sin\left(\frac{3\pi}{2}\right), \sin(0)\right]^T\\
&p_{11}^d = r.\left[\cos\left(-\frac{\pi}{4}\right)\cos\left(\frac{2\pi}{3}\right), \cos\left(-\frac{\pi}{4}\right)\sin\left(\frac{2\pi}{3}\right), \sin\left(-\frac{\pi}{4}\right)\right]^T\\
&p_{12}^d = r.\left[\cos\left(-\frac{\pi}{4}\right)\cos\left(\frac{4\pi}{3}\right), \cos\left(-\frac{\pi}{4}\right)\sin\left(\frac{4\pi}{3}\right), \sin\left(-\frac{\pi}{4}\right)\right]^T\\
&p_{13}^d = r.\left[\cos\left(-\frac{\pi}{4}\right)\cos(2\pi), \cos\left(-\frac{\pi}{4}\right)\sin(2\pi), \sin\left(-\frac{\pi}{4}\right)\right]^T
\end{aligned}
\tag{7.12}
$$

where the radius of the sphere is $r = 2$ m.

The initial attitude of each aerial vehicle is generated randomly, on condition that the initial communication network is leader-follower connected. Without loss of generality, the initial position of each aerial vehicle is shown in Fig. 7.6a, which is expressed by the green star point. According to the initial position of each agent, the initial communication network of the escort system is given as

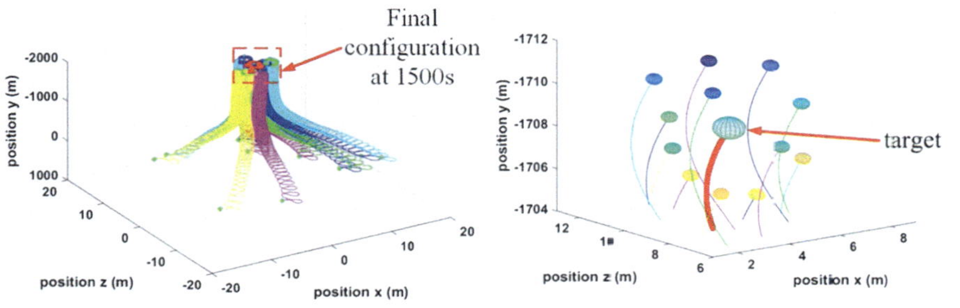

(a) (b)

Fig. 7.6 Which means that the connectivity of the communication topology is preserved from 0 to 1500 s. Thus, conclusion 1 of Corollary 1 is proved. **a** Trajectory. **b** Final configuration at time 1500 s

$$A(0) = \begin{bmatrix} 1&0&0&0&0&0&0&0&0&0&0&0&0 \\ 1&0&1&0&1&1&1&1&0&0&1&1&1 \\ 1&1&0&1&1&1&1&1&1&1&1&1&1 \\ 1&0&1&0&1&1&0&1&1&1&1&1&1 \\ 1&1&1&1&0&1&0&1&0&0&1&1&1 \\ 1&1&1&1&1&0&1&1&1&1&1&1&1 \\ 1&1&1&0&0&1&0&1&1&1&1&1&1 \\ 1&1&1&1&1&1&1&0&1&1&1&1&1 \\ 1&0&1&1&0&1&1&1&0&1&1&1&1 \\ 1&0&1&1&0&1&1&1&1&0&1&1&1 \\ 1&1&1&1&1&1&1&1&1&1&0&1&1 \\ 1&1&1&1&1&1&1&1&1&1&1&0&1 \\ 1&1&1&1&1&1&1&1&1&1&1&1&0 \end{bmatrix} \tag{7.13}$$

Let the simulation time be $T_f = 1500$ s. Then, the simulation results of the escort system (7.3) applying the control protocols (7.6) and (7.10) are shown in Figs. 7.6, 7.7 and 7.8. The final communication network of the escort system is given as

$$A(T_f) = \begin{bmatrix} 1&0&0&0&0&0&0&0&0&0&0&0&0 \\ 1&0&1&1&1&1&1&1&1&1&1&1&1 \\ 1&1&0&1&1&1&1&1&1&1&1&1&1 \\ 1&1&1&0&1&1&1&1&1&1&1&1&1 \\ 1&1&1&1&0&1&1&1&1&1&1&1&1 \\ 1&1&1&1&1&0&1&1&1&1&1&1&1 \\ 1&1&1&1&1&1&0&1&1&1&1&1&1 \\ 1&1&1&1&1&1&1&0&1&1&1&1&1 \\ 1&1&1&1&1&1&1&1&0&1&1&1&1 \\ 1&1&1&1&1&1&1&1&1&0&1&1&1 \\ 1&1&1&1&1&1&1&1&1&1&0&1&1 \\ 1&1&1&1&1&1&1&1&1&1&1&0&1 \\ 1&1&1&1&1&1&1&1&1&1&1&1&0 \end{bmatrix} \tag{7.14}$$

which means that the connectivity of the communication topology is preserved from 0 to 1500 s. Thus, conclusion 1 of Corollary 7.1 is proved.

Figure 7.7a shows the distances between any two individuals of the air system. Figure 7.7b shows that the distance between the target aerial vehicle and each miniature UAV. Since that these distances are larger than the safe distance between any two aerial vehicles, the conclusion 2 of Corollary 7.1 is proved.

Figure 7.8 gives the differences of the yaw angle, the pitch angle, the thrusting speed, and the rotational speed between the target aerial vehicle and each miniature UAV. These figures show that the attitude angles and the speeds of each miniature UAV converge to the same with those of the target aerial vehicle. Thus, conclusion 3 of Corollary 7.1 is proved.

In Fig. 7.6a, the colorful ball is the final position of each aerial vehicle at time $t = 1500$ s. Figure 7.6b gives a more clear exhibition on the final configuration of the

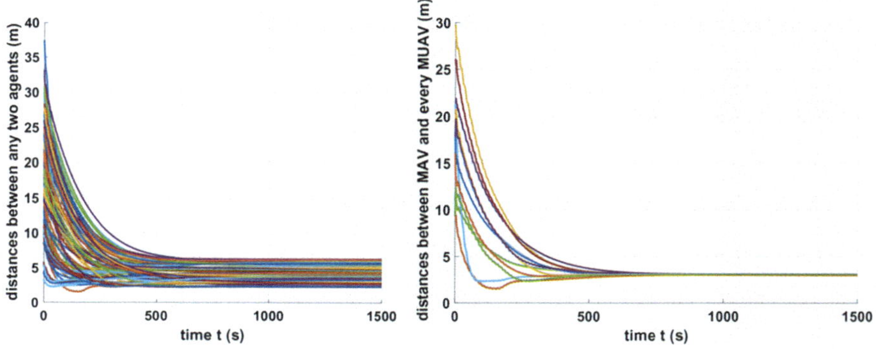

(a) (b)

Fig. 7.7 Simulation results of the distances between any two agents. **a** Distances between any two agents. **b** Distances between target aerial vehicle and every miniature UAV.

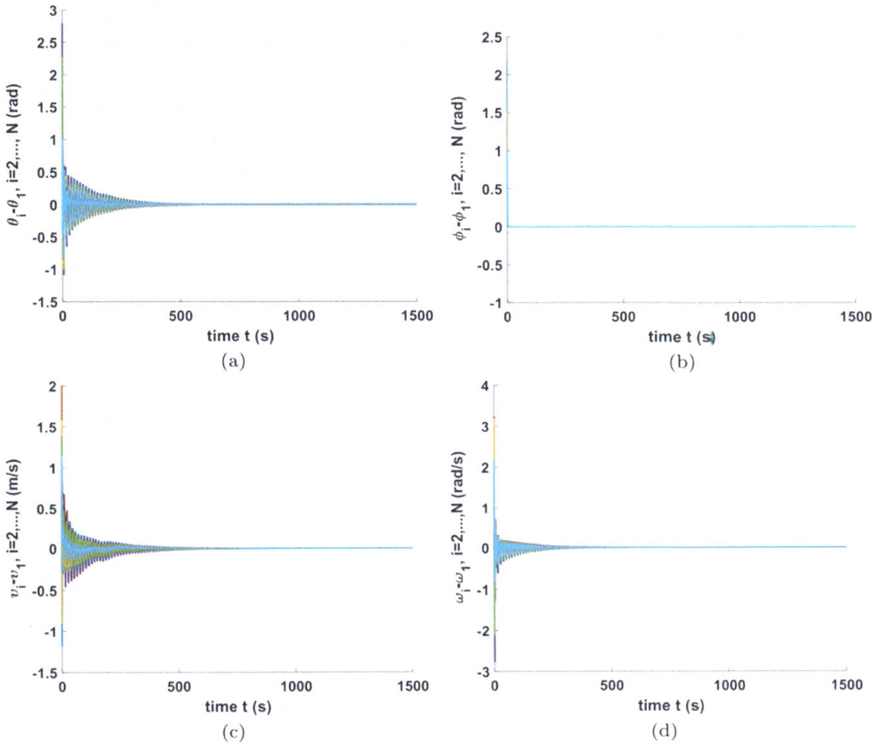

Fig. 7.8. Simulation results of the attitude angles and the speeds of one target aerial vehicle and twelve miniature UAVs. **a** Differences of yaw angle versus time. **b** Differences of pitch angle. **c** Differences of thrusting speed. **d** Differences of rotational speed.

thirteen agents at time $t = 1500s$, which is a three-dimensional sphere formation. Figure 7.7a indicates that the distances between any two aerial vehicles asymptotically become stable, which implies that the formation of the thirteen aerial vehicles can be maintained. Thus, conclusion 4 of Corollary 7.1 is proved.

(2) **Herringbone Aerial Escort Task**

Herringbone formation is inspired from the wild gooses flying in compact formations during migrations in order to save energy. Thus, herringbone aerial escort becomes another special formation aerial escort task that would be studied further in this section. The herringbone escort mainly aims to influence the enemy's judgement to the position of the target aerial vehicle. These miniature UAVs act as decoys during the escort task. Therefore, the herringbone aerial escort task is to build such a formation that whose center is not at the position of the target aerial vehicle.

Similarly, the initial attitudes of one target aerial vehicle and four miniature UAVs are generated randomly, on condition that the initial communication network is

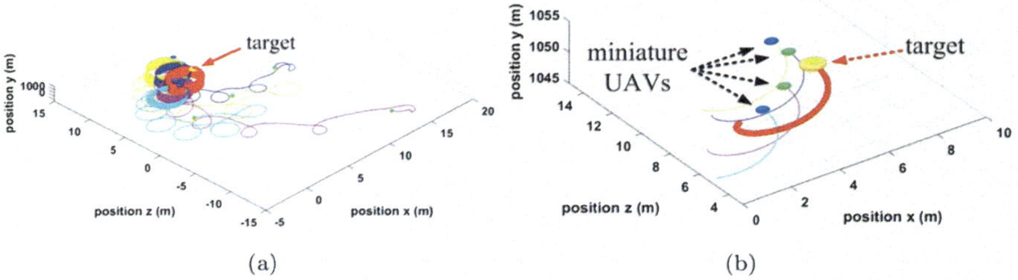

Fig. 7.9. Simulation results of the herringbone formation's trajectories of one target aerial vehicle and four miniature UAVs. **a** Trajectory. **b** Final configuration at time 800 s

leader-follower connected. The simulation results of the escort system (7.3) applying the control protocols (7.6) and (7.10) to imitate the three-dimensional flight of the wild gooses are shown in Figs. 7.9 and 7.10. As shown in Fig. 7.9, the target aerial vehicle flies like the leading wild goose at the front of the herringbone formation, and four miniature UAVs fly on both sides of the target aerial vehicle and follows the flight of the target aerial vehicle. Figure 7.10 shows that the differences of the yaw angle, the pitch angle, the thrusting speed, and the rotational speed between the target aerial vehicle and each miniature UAV versus time. It is obvious that the attitude angles and the speeds of each miniature UAVs converge to the same with those of the target aerial vehicle. Besides, as shown in Fig. 7.11, the distances between any two aerial vehicles also converge to be stable, which means that the herringbone formation of one target aerial vehicle and four miniature UAVs can be maintained without disturbance. So far, it is concluded that the control protocols (7.6) and (7.10) are effective for the escort system (7.3) to imitate the three-dimensional flight of the wild gooses.

7.3 Summary

Several kinds of three-dimensional aerial escort tasks have been investigated by using networked miniature UAVs for the sake of safe flight of one target aerial vehicle. The aerial escort task involved in this paper is simplified as a three-dimensional formation control problem of multi-agent system with leader-follower structure. Therein, the target aerial vehicle is taken as the leader, and the miniature UAVs are considered as the followers. Each aerial vehicle is modeled as an extended second-order unicycle with limited communication capability. A flocking algorithm is proposed to enable the miniature UAVs asymptotically track the target aerial vehicle's velocity and form a stable cohesive or geometrical configuration together with the target aerial vehicle, provided that the initial communication network is leader-follower

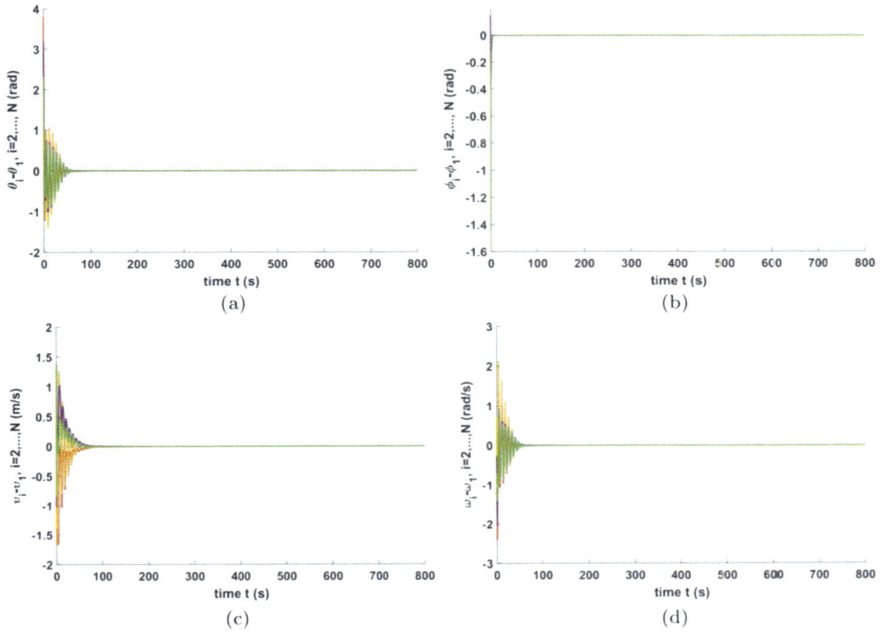

Fig. 7.10. Simulation results of the herringbone formation of one target aerial vehicle and four miniature UAVs. **a** Differences of yaw angle. **b** Differences of pitch angle. **c** Differences of thrusting speed. **d** Differences of rotational speed

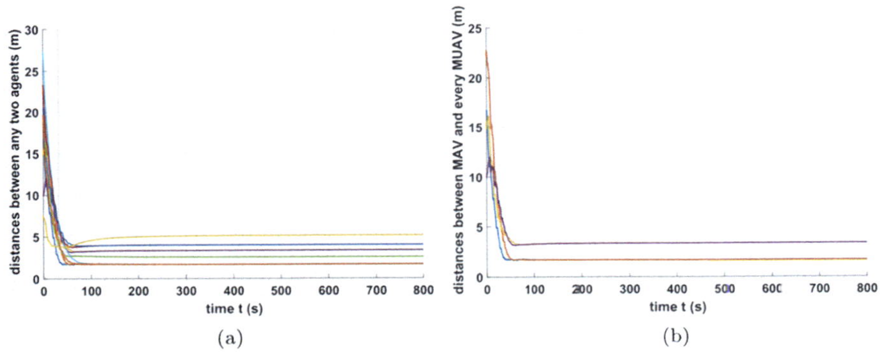

Fig. 7.11. Simulation results of the herringbone formation of one target aerial vehicle and four miniature UAVs. **a** Distances between any two agents. **b** Distances between target aerial vehicle and every miniature UAV

connected. Besides, no collision happens between any two aerial vehicles. The robustness of the escort system is also illustrated. The simulation results of the cohesive escort task, the sphere escort task, and the herringbone escort task are all given to further validate the functionality of the proposed algorithm. In the next step, the aerial escort tasks for more than one targets are going to be studied in further, and the optimization of the proposed control algorithm also will be discussed in order to satisfy the requirements for actual combat.

References

1. Charlton, J. (2015). *The Navy's swarming locust drones*. Military Aviation.
2. Quick, D. (2015). *US Navy goes tubular with autonomous swarming UAV demonstrations*. Office of Naval Research.
3. Jadbabaie, A., Lin, J., & Morse, A. S. (2003). Coordination of groups of mobile autonomous agents using nearest neighbor rules. *IEEE Trans actions on Automatic Control, 48*(6), 988–1001.
4. Loria, A., Dasdemir, J., & Alvarez Jarquin, N. (2016). Leader-follower for mation and tracking control of mobile robots along straight paths. *IEEE Transactions on Control Systems Technology, 24*(2), 727–732.
5. Mariottini, G. L., Morbidi, F., Prattichizzo, D., Vander, V. N., Michael, N., Pappas, G., & Daniilidis, K. (2009). Vision-based localization for leader-follower formation control. *IEEE Transactions on Robotics, 25*(6), 1431–1438.
6. Balch, T., & Arkin, R. C. (1998). Behavior-based formation control for multirobot teams. formation control for multirobot teams. *IEEE Transactions on Robotics and Automation, 14*(6), 926–939.
7. Monteiro, S., & Bicho, E. (2002). A dynamical systems approach to behavior-based formation control. In *Proceedings 2002 IEEE International Conference on Robotics and Automation*, Washington, DC, USA.
8. Shao, J., Wang, L., & Yu, J. (2008). Development of an artificial fish-like robot and its application in cooperative transportation. *Control Engineering Practice, 16*(5), 569–584.
9. Chen, X., & Jia, Y. (2015). Adaptive leader-follower formation control of non-holonomic mobile robots using active vision. *Control Theory & Applications IET, 9*(8), 1302–1311.
10. Low, C. B. (2011). A dynamic virtual structure formation control for fixed-wing UAVs. In *9th IEEE International Conference on Control and Automation (ICCA)*, 19–21 Dec 2011, Santiago, Chile. IEEE.
11. Ren, W., & Beard, R. W. (2004). Formation feedback control for multiple spacecraft via virtual structures. *IEE Proceedings—Control Theory and Applications, 151*(3), 357–368.
12. Li, R., Zhang, L., Han, L., & Wang, J. (2017). Multiple vehicle formation control based on robust adaptive control algorithm. *IEEE Intelligent Transportation Systems Magazine, 9*(2), 41–51.
13. Park, B. S., Jin, B. P., & Yoon, H. C. (2011). Adaptive formation control of electrically driven nonholonomic mobile robots with limited information. *IEEE Transactions on Systems, Man, and Cybernetics, Part B (Cybernetics), 41*(4), 1061–1075.
14. Peng, Z., Wang, D., & Hu, X. (2011). Robust adaptive formation control of under actuated autonomous surface vehicles with uncertain dynamics. *IET Control Theory & Applications, 5*(12), 1378–1387.
15. Sarangapani, J., & Dierks, T. A. (2007). *Neural network control of robot formations using rise feedback*. IEEE.
16. Defoort, M., Floquet, T., Kokosy, A., & Perruquetti, W. (2008). Sliding mode formation control for cooperative autonomous mobile robots. *IEEE Transactions on Industrial Electronics, 55*(11), 3944–3953.

17. Xiao, H., Li, Z., & Chen, C. (2016). Formation control of leader-follower mobile robots systems using model predictive control based on neuro dynamics optimization. *IEEE Transactions on Industrial Electronics, 63*(9), 1–1.
18. Yu, X., & Liu, L. (2016). Distributed formation control of nonholonomic vehicles subject to velocity constraints. *IEEE Transactions on Industrial Electronics, 63*(2), 1289–1298.
19. Guan, Y., & Wang, L. (2017). Structural controllability of multi-agent systems with absolute protocol under fixed and switching topologies. *Science China (Information Sciences), 60*(9), 226–240.
20. Lu, Z., Zhang, L., & Wang, L. (2019). Controllability analysis of multiagent systems with switching topology over finite fields. *Science China (Information Sciences), 62*(1), 80–94.
21. Chen, L., & Ma, B. (2015). A nonlinear formation control of wheeled mobile robots with virtual structure approach. In *34th Chinese Control Conference*, Hangzhou, China (pp. 1080–1085).
22. Wang, H., Guo, D., Liang, X., Chen, W., Hu, G., & Leang, K. K. (2016). Adaptive vision-based leader-follower formation control of mobile robots. *IEEE Transactions on Industrial Electronics, 64*, 2893–2902.
23. Jia, Y., & Wang, L. (2015). Leader-follower flocking of multiple robotic fish. *IEEE/ASME Transactions on Mechatronics, 20*(3), 1372–1383.
24. Cucker, F., & Smale, S. (2007). Emergent behavior in flocks. *IEEE Transactions on Automatic Control, 52*(5), 852–862.
25. Ferrante, E., Turgut, A. E., Huepe, C., Stranieri, A., Pinciroli, C., & Dorigo, M. (2012). Self-organized flocking with a mobile robot swarm: a novel motion control method. *Adaptive Behavior, 20*(6), 460–477.
26. Olfati-Saber, R. (2006). Flocking for multi-agent dynamic systems: Algorithms and theory. *IEEE Transactions on Automatic Control, 51*(3), 401–420
27. Reynolds, C. W. (1987). Flocks, herds and schools: A distributed behavioral model. *ACM SIGGRAPH Computer Graphics, 21*(4), 25–34.
28. Tu, C., Yang, C., Liu, Z., & Sun, M. (2017). August). Network representation learning: An overview. *SCIENTIA SINICA Informationis, 47*, 980–996.

Chapter 8
Optimization Strategies for Distributed Coordination Control Systems

In the previous chapters, we focused on the mechanisms underlying biological swarm behavior and how to leverage these mechanisms to design distributed cooperative control strategies for multi-agent systems. These strategies aim to achieve swarm control for either formation or geometric tasks. In practice, however, it is not just about completing tasks but about how effectively and efficiently those tasks are completed. For instance, can the specified swarm task be completed more quickly? Can it be achieved in a more stable state?

This chapter provides an initial exploration into optimizing the design of distributed cooperative control systems, offering preliminary insights. We analyze and experiment with various aspects, such as system parameters (e.g., communication network weights), communication radius, system architecture, and individual perspectives. These analyses have led to some interesting conclusions. We believe these efforts contribute to a deeper understanding of swarm concepts and behaviors, laying a solid theoretical foundation for the practical application of swarm strategies.

Furthermore, to better facilitate collaborative research outcomes, we have established a unified platform for studying swarm behavior (details are covered in Sect. 8.3). This platform allows for the abstraction of various swarm phenomena into mathematical models for quantitative and qualitative analysis and comparison. We look forward to the continuous emergence of more classic swarm models.

8.1 Optimization of System Parameters

Have you ever seen flocks of birds maneuvering swiftly and precisely as a single entity at sunset? The memorable synchronization phenomenon definitely stops you in your tracks [1, 2]. Nature uses its magical creativity to produce an illusion of an enormous and dynamical sand painting with thousands of birds. Unmanned aerial vehicle (UAV) flocking is such an artificial system that exploits these superior swarm

Y. Jia, *Distributed Coordination Control System*,
https://doi.org/10.1007/978-981-96-0365-7_8

intelligences of flocks of birds [3]. Realization of flocking behavior of multi-agent system has been eagerly required for civil and military missions for its remarkable advantages in improving task effectiveness, minimizing the probability of detection, and achieving task execution flexibility.

Fixed-wing UAVs are commonly the first choice for implementing flocking tasks because of their remarkable altitude advantage, which maximizes the field of vision. This chapter applies the complete six-degrees-of-freedom model, including 12 variables, to draw the kinematic and dynamical characteristics of the fixed-wing UAV. Based on the proposed model and the weighted metric interaction mechanism, a distributed cooperation approach is designed to force groups of fixed-wing UAVs to fly like flocks of birds. Three essential attributes, separation, alignment, and cohesion, are all proved to be satisfied. Furthermore, 4 constraint conditions are considered in this system, which are tactfully considered as evaluation criteria to designate the communication network's coupling strength. Finally, simulation results validate the effectiveness of the proposed algorithm for the steady flight task of variable-altitude flocking and the steady flight task of invariable-altitude flocking.

8.1.1 Introduction

Unmanned aerial vehicles (UAVs) own natural altitude superiority in civil and military missions. Therefore, UAV flocking has attracted great attention from physicists and engineers, as well as from military scientists, who believe that UAV flocking will play an irreplaceable role in future wars, because of its remarkable superiority in improving combat effectiveness, extending UAV survivability, and achieving task execution flexibility [4].

Various studies on UAV flocking have been conducted, resulting in recent advancements in modelling [5–7], communication networks [8–11], cooperation approaches [12–15], simulation and experimental validation [19–23], and task-based design [16–18]. Agent-based modeling and simulation approaches are capable of discovering the intuitive paradigm of swarms and, in particular, offer tremendous explanatory capability that leads to unpredictable, non-linear results [7].

Stability Analysis of flocking is another topic that cannot be ignored. Eigenvalues method is a useful tool for the stability analysis of the linear flocking system, such as Cucker-Smale model [24]. In general, the whole system will reach a global stability on condition that the feedback gain coefficient falls in a specific range [24]. Meanwhile, the stability of nonlinear flocking system is usually analyzed by using LaSalle-Krasovskii invariance principle [25], which can be considered as an extension of Lyapunov function method. According to LaSalle-Krasovskii invariance principle, the whole system will reach a global stability on condition that the initial interaction network of the system is connected [26].

As we all know, quadrotor UAVs have versatile kinetic characteristics, including spot hover, vertical take-off and landing, and backward flying. These remarkable capabilities make quadrotor UAVs cardinal flying platforms for exploring the flocking problems [18, 27]. However, fixed-wing UAVs are more suitable for the battlefield environment because of their faster speed, longer endurance, and even better stealth performance. Therefore, this paper focuses on the fixed-wing UAV flocking problem, which has more significant challenges than that of quadrotor UAVs. First, the fixed-wing UAV's aerodynamic characteristics are more complex than those of quadrotor UAVs. Second, fixed-wing UAVs lack the spot hover capability, the stop-and-wait path deconfliction algorithms suitable for quadrotors are thus not applicable to fixed-wing UAVs. Third, a forward speed is necessary for fixed-wing UAV flight [28].

In this subsection, fixed-wing UAV is presented as a three-dimensional kinematic and dynamic model with a complete set of 12 variables on the assumption that each fixed-wing UAV flies under bounded attitude angle and bounded velocity. The interaction network of the fixed-wing UAV system employs the nearest neighbor rules [29]. One weighted and time-variant communication topology is considered, whose coupling strength denotes the contribution of the neighbor of each fixed-wing UAV. Provided that different neighbors offer different contributions to the same fixed-wing UAV, a kind of adaptive strategy was adopted to compute the weight of the coupling strength depending on the relative velocities and relative distances between any two individuals [30]. In this paper, a simple and effective method is proposed to obtain distance-related coupling strengths, which seems more closed to real situations.

Inspired by the collective behaviors of social animals, such as birds, a decentralized control protocol is proposed on the basis of a consensus algorithm that leads to consistent behavior among fixed-wing UAVs. The artificial field method is also adopted by the protocol to ensure collision avoidance, connectivity preservation, and cohesive flocking. Additionally, the protocol's last section offsets the gyroscopic effect generated by the fixed-wing UAV's rotational motion. Provided that the initial interaction network of the system is an undirected connected graph, the stability of flocking is analyzed. The parameter selection problem that designates a value to the coupling strength of the communication network is introduced by means of taking the constraint conditions as evaluation criteria, in case the coupling strength between any two fixed-wing UAVs is distance-related. The simulation is carried out to choose a proper coupling strength and verify the theoretical results of the flocking behaviors of fixed-wing UAVs.

The most significant contribution of this section is the first attempt to combine control theory with statistical methods to address distributed cooperative control problems. Specifically, the approach involves two steps: first, using control theory tools to perform a rigorous stability analysis of the system; and then, applying statistical methods to optimize the communication topology strength coefficients.

8.1.2 Modeling of UAV

The system under consideration consists of N fixed-wing UAVs with three-dimensional flying capabilities. Each fixed-wing UAV is taken as a sphere with radius R. $2R$ is the safe distance between any two fixed-wing UAVs. To save space, we omit the time-dependence in the following sections. For example, we use x_i instead of $x_i(t)$.

It is well-known that the fixed-wing UAV has complex aerodynamic characteristics. Thus, several assumption conditions should be given to simplify the difficulty of the modeling process. First, the flat earth hypothesis is adopted for the modeling process. The curvature of the earth is ignored and the earth coordinate system is taken as inertial. Second, each fixed-wing UAV is rigid without consideration of its elastic deformation while flying at high altitudes.

Based on the proposed assumptions and without loss of generality, the fixed-wing UAV i ($i = 1, \ldots, N$) can be modeled by the following equations [31]:

$$
\begin{aligned}
\dot{x}_i &= u_i \cos\theta_i \cos\psi_i + \upsilon_i(\sin\phi_i \sin\theta_i \cos\psi_i - \cos\phi_i \sin\psi_i) \\
&\quad + \omega_i(\sin\phi_i \sin\psi_i + \cos\phi_i \sin\theta_i \cos\psi_i) \\
\dot{y}_i &= u_i \cos\theta_i \sin\psi_i + \upsilon_i(\sin\phi_i \sin\theta_i \sin\psi_i + \cos\phi_i \cos\psi_i) \\
&\quad + \omega_i(-\sin\phi_i \cos\psi_i + \cos\phi_i \sin\theta_i \sin\psi_i) \\
\dot{z}_i &= u_i \sin\theta_i - \upsilon_i \sin\phi_i \cos\theta_i - \omega_i \cos\phi_i \cos\theta_i \\
\dot{\phi}_i &= p_i + (r_i \cos\phi_i + q_i \sin\phi_i)\tan\theta_i \\
\dot{\theta}_i &= q_i \cos\phi_i - r_i \sin\phi_i \\
\dot{\psi}_i &= \frac{1}{\cos\theta_i}(r_i \cos\phi_i + q_i \sin\phi_i) \\
\dot{u}_i &= \upsilon_i r_i - \omega_i q_i - g_i \sin\theta_i + \frac{F_{xi}}{m_i} \\
\dot{\upsilon}_i &= -u_i r_i + \omega_i p_i + g_i \cos\theta_i \sin\phi_i + \frac{F_{yi}}{m_i} \\
\dot{\omega}_i &= u_i q_i - \upsilon_i p_i + g_i \cos\theta_i \cos\phi_i + \frac{F_{zi}}{m_i} \\
\dot{p}_i &= (c_{1i} r_i + c_{2i} p_i) q_i + c_{3i}\overline{L}_i + c_{4i}\overline{N}_i \\
\dot{q}_i &= c_{5i} p_i r_i - c_{6i}(p_i^2 - r_i^2) + c_{7i}\overline{M}_i \\
\dot{r}_i &= (c_{8i} p_i - c_{2i} r_i) q_i + c_{4i}\overline{L}_i + c_{9i}\overline{N}_i
\end{aligned}
\tag{8.1}
$$

where x_i, y_i, z_i respectively denote the x-axis position coordinate, the y-axis position coordinate, and the z-axis position coordinate of the fixed-wing UAV i at time t; ϕ_i, θ_i, ψ_i respectively denote the roll angle, the pitch angle, and the yaw angle of the fixed-wing UAV i at time t; u_i, v_i, ω_i respectively denote the x-axis speed component, the y-axis speed component, and the z-axis speed component of the fixed-wing UAV i at time t; p_i, q_i, r_i respectively denote the roll angle speed, the pitch angle speed, and the yaw angle speed of the fixed-wing UAV i at time t; F_{xi}, F_{yi}, F_{zi} respectively denote the x-axis component, the y-axis component, and the z-axis component of the force acting on the fixed-wing UAV i at time t; $\overline{L}_i, \overline{M}_i, \overline{N}_i$ respectively denote the roll torque, the pitch torque, and the yaw torque acting on the fixed-wing UAV i at time t; m_i is the quality of the fixed-wing UAV i at time t; and g_i is the acceleration of gravity of the fixed-wing UAV i at time t. Therein,

$$
\begin{aligned}
&\dot{c}_{1i} = \frac{(I_{yi} - I_{zi})I_{zi} - I_{xzi}^2}{\sum_i}, c_{2i} = \frac{(I_{xi} - I_{yi} + I_{zi})I_{xzi}}{\sum_i} \\
&\dot{c}_{3i} = \frac{I_{zi}}{\sum_i}, c_{4i} = \frac{I_{xzi}}{\sum_i}, c_{5i} = \frac{I_{zi} - I_{xi}}{I_{yi}}. c_{6i} = \frac{I_{xzi}}{I_{yi}} \\
&\dot{c}_{7i} = \frac{1}{I_{yi}}, c_{8i} = \frac{I_{xi}(I_{xi} - I_{yi}) + I_{xzi}^2}{\sum_i}, c_{9i} = \frac{I_{xi}}{\sum_i}
\end{aligned} \tag{8.2}
$$

where.$I_{xi} = \int (y_i^2 + z_i^2)dm_i$. denotes the x-axis rotational inertia of the fixed-wing UAV i at time t, $I_{yi} = \int (x_i^2 + z_i^2)dm_i$ denotes the y-axis rotational inertia of the fixed-wing UAV i at time t, $I_{zi} = \int (y_i^2 + x_i^2)dm_i$ denotes the z-axis rotational inertia of the fixed-wing UAV i at time t, $I_{xzi} = \int x_i z_i dm_i$ denotes the product of inertia, and $\sum_i I_{xi} I_{zi} - I_{xzi}^2$.

In particular, individual differences exist among the fixed-wing UAVs: that is to say, m_i may not be equal to m_j for $i, j = 1, ..., N$, $i \neq j$. Moreover, the value of m_i may change versus time because of fuel consumption. Note that g_i may change with the altitude of the fixed-wing UAV i. We leave out the difference in the flying altitude of each fixed-wing UAV in the colony; thus, we have $g_i = g_j$ for $i, j = 1, ..., N$.

Define that $P_i = [x_i, y_i, z_i]^T$ as the position vector, $\Theta_i = [\phi_i, \theta_i, \psi_i]^T$ as the attitude-angle vector, $Q_i(t) = [u_i, v_i, \omega_i]^T$ as the speed vector, and $S_i = [p_i, q_i, r_i]^T$ as the attitude-angle speed vector. Then, the equations in (8.1) can be merged into the following matrix form:

$$
\begin{aligned}
&\dot{P}_i = H_i^T Q_i \\
&\dot{Q}_i = M_i Q_i + G_i g_i + \frac{1}{m_i} U_i^F \\
&\dot{\Theta}_i = D_i^T S_i \\
&\dot{S}_i = K_i U_i^M + B_i
\end{aligned} \tag{8.3}
$$

where $G_i = [-\sin\theta_i, \cos\theta_i \sin\phi_i, \cos\theta_i \cos\phi_i]^T$, $H_i = \left[H_i^1, H_i^2, H_i^3\right]$,

$$
M_i = \begin{bmatrix} 0 & r_i & -q_i \\ -r_i & 0 & p_i \\ q_i & -p_i & 0 \end{bmatrix} \tag{8.4}
$$

$$
D_i = \begin{bmatrix} 1 & 0 & 0 \\ \sin\phi_i \tan\theta_i & \cos\phi_i & \frac{\sin\phi_i}{\cos\theta_i} \\ \cos\phi_i \tan\theta_i & -\sin\phi_i & \frac{\cos\phi_i}{\cos\theta_i} \end{bmatrix} \tag{8.5}
$$

$$K_i = \begin{bmatrix} c_{3i} & 0 & c_{4i} \\ 0 & c_{7i} & 0 \\ c_{4i} & 0 & c_{9i} \end{bmatrix} \tag{8.6}$$

and,

$$B_i = \begin{bmatrix} (c_{1i}r_i + c_{2i}p_i)q_i \\ c_{5i}p_ir_i - c_{6i}(p_i^2 - r_i^2) \\ (c_{8i}p_i + c_{2i}r_i)q_i \end{bmatrix} \tag{8.7}$$

Therein,

$$H_i^1(t) = \begin{bmatrix} \cos\theta_i \cos\psi_i \\ \sin\phi_i \sin\theta_i \cos\psi_i - \cos\phi_i \sin\psi_i \\ \sin\phi_i \sin\psi_i + \cos\phi_i \sin\theta_i \cos\psi_i \end{bmatrix}, \tag{8.8}$$

$$H_i^2(t) = \begin{bmatrix} \cos\theta_i \sin\psi_i \\ \sin\phi_i \sin\theta_i \sin\psi_i + \cos\phi_i \cos\psi_i \\ -\sin\phi_i \cos\psi_i + \cos\phi_i \sin\theta_i \sin\psi_i \end{bmatrix}, \tag{8.9}$$

and

$$H_i^3(t) = \begin{bmatrix} \sin\theta_i \\ -\sin\phi_i \cos\theta_i \\ -\cos\phi_i \cos\theta_i \end{bmatrix}. \tag{8.10}$$

Note that the force acting on the ith fixed-wing UAV $U_i^F = [F_{xi}, F_{yi}, F_{zi}]^T$ is obtained by the combination of the aerodynamic force (including the lift force L_i, drag force D_i, and lateral force Y_i) and the thrust of the engine T_i; that is,

$$\vec{i}F_{xi} + \vec{j}F_{yi} + \vec{k}F_{zi} = L_i + D_i + Y_i + T_i \tag{8.11}$$

where $L_i = C_{Li}Q_iS_{wi}$, $D_i = C_{Di}Q_iS_{wi}$, $Y_i = C_{Yi}Q_iS_{wi}$, and $T_i = C_{Ti}Q_iS_{Di}\delta_{Ti}$. $\vec{i}, \vec{j}, \vec{k}$ are respectively the unit vector of x-axis, y-axis, and z-axis. C_{Li}, C_{Di}, C_{Yi}, C_{Ti} are respectively the lift force coefficient, the drag force coefficient, the lateral force coefficient, and the thrust force coefficient of the UAV \$i\$. Dynamic pressure $Q_i = \frac{1}{2}\rho_i V_i^2$, where airspeed $V_i = (u_i^2 + v_i^2 + \omega_i^2)^{0.5}$, angle of attack $\alpha = \arctan\frac{\omega}{u}$, angle of sideslip $\beta = \arctan\frac{v}{V}$, and air density $\rho_i = \rho_i^0(1 - 0.703 \times 10^{-5} \times z_i)^{4.14}$. ρ_i^0 is the air density at the initial time. S_{wi} is the reference area of the wing of the UAV i, S_{Di} is the blade area of the UAV i, and δ_{Ti} is the open degree of the engine accelerator.

The total torque $U_i^M = [\overline{L}_i, \overline{M}_i, \overline{N}_i]^T$ can also be obtained by

$$\begin{aligned}\overline{L}_i &= C_{li}Q_iS_{wi}b_i\delta_{ai}\\ \overline{M}_i &= C_{mi}Q_iS_{wi}C_{Ai}\delta_{ei}\\ \overline{N}_i &= C_{ni}Q_iS_{wi}b_i\delta_{ri},\end{aligned} \tag{8.12}$$

where C_{li}, C_{mi}, C_{ni} are, respectively, the roll torque coefficient, pitch torque coefficient, and yaw torque coefficient of UAV i; C_{Ai} is the average aerodynamic chord length; b_i is the length of the wingspan of UAV i; and δ_{ai}, δ_{ei}, δ_{ri} are, respectively, the aileron's deflection angle, the elevator's deflection angle, and the rudder's deflection angle.

The above equations give the nonlinear relationship between the state vector $FS_i = [P_i^T, \Theta_i^T, Q_i^T, S_i^T]^T$ and the control input vector $[\delta_{ai}, \delta_{ei}, \delta_{ri}, \delta_{Ti}]^T$ of the fixed-wing UAV i. However, (8.11) and (8.12) are closely related to the kind of fixed-wing UAV chosen. To endow the control algorithm with more common applications, we just use U_i^F and U_i^M as the control input instead of $[\delta_{ai}, \delta_{ei}, \delta_{ri}, \delta_{Ti}]^T$, which can be obtained easily from FS_i, U_i^F, and U_i^M according to (8.11) and (8.12) at each time t.

8.1.3 Interaction Mechanism

Because of its limited communication capability, each fixed-wing UAV is supposed to be able to interact only with its neighbors. The neighbor relationship is signified by a dynamic undirected graph ξ consisting of a node set $v = \{1, 2, ..., N\}$ and a time-varying edge set $\varepsilon \in v \times v$. Therein, v denotes the fixed-wing UAVs and ε denotes the neighbor relationships between any two fixed-wing UAVs. If UAV i and UAV j are neighbors, we have $(j, i) \in \varepsilon(t)$. Note that v is finite and nonempty.

8.1.3.1 Metric Neighbor Relationship

As mentioned above, each UAV can only interact with its neighbors in a restricted vicinity [32]. The neighbor set of fixed-wing UAV i at time t is denoted by $N_i(t)$.

First, the initial neighbor set of UAV i at time t is defined as [14]

$$N_i(0) = \{j \big| ||P_i(0) - P_j(0)|| < D, i, j = 1, ..., N, j \neq i\} \tag{8.13}$$

where $||\cdot||$ is the Euclidean norm and $D > 2R$ is the interaction radius.

Second, because the distance between any two UAVs will change before the system reaches stability, the neighbor relationship of two UAVs also changes. Suppose that the communication network $\xi(t)$ is supposed to switch at t_p, $p = 1, 2, \ldots$. To preserve the connectivity of the interaction network $\xi(t)$, the hysteresis adding new edges to the network is introduced [14], such that

(1) if $(i, j) \in \varepsilon(t^-)$ and $||P_i(t) - P_j(t)|| < D - \varepsilon_0$, $\varepsilon_0 \in (0, D)$ is a given constant, $i = 1, ..., N$, then $(i, j) \in \varepsilon(t)$, for $t > 0$;

(2) if $||P_i(t) - P_j(t)|| \geq D$, then $(i, j) \notin \varepsilon(t)$, for $t > 0$.

Thus, $\xi(t)$ is a fixed graph in each nonempty, bounded, and contiguous time interval $[t_r, t_{r+1})$, where $r = 0, ..., N$.

8.1.3.2 Metric Interaction Network

To clarify the neighbor relationship, the coupling configuration of the network $\xi(t)$ is presented by the following adjacency matrix $A_N = [w_{ij}]_{N\times N} \in M_N(R)$.

$$w_{ij}(t) = \begin{cases} 1 & (j, i) \in \varepsilon(t) \\ 0 & otherwise \end{cases} \tag{8.14}$$

It is easy to believe that different neighbors of one UAV should have different efforts on deciding the UAV's behavior, based on which we introduce a new parameter k_{ij}. k_{ij} denotes the coupling strength of neighbors i and j. Then, the adjacency matrix of the weighted network $\xi(t)$ is given by

$$B(t) = \begin{bmatrix} k_{11}w_{11} & k_{12}w_{12} & \cdots & k_{1N}w_{1N} \\ k_{21}w_{21} & k_{22}w_{22} & \cdots & k_{2N}w_{2N} \\ \vdots & \vdots & \ddots & \vdots \\ k_{N1}w_{N1} & k_{N2}w_{N2} & \cdots & k_{NN}w_{NN} \end{bmatrix} \tag{8.15}$$

where $k_{ij} > 0$ and $k_{ij} = k_{ji}$, $i, j \in v$. Moreover, the coupling strength k_{ij} is related to the distance between UAV i and j. Thus, the definition of k_{ij} is given by

$$k_{ij} = \frac{k}{\|P_i - P_j\|} \tag{8.16}$$

where $k > 0$ is a positive constant. For UAV j, the nearer from the UAV i, the more effect on UAV i.

Furthermore, the Laplacian matrix $L_N = [l_{ij}]_{N\times N} \in M_N(R)$ of the network $\xi(t)$ is proposed as

$$l_{ij}(t) = \begin{cases} -k_{ij}w_{ij}, \ i \neq j \\ \sum\limits_{k=1,k\neq i}^{N} k_{ik}w_{ik}, \ i \neq j \end{cases} \tag{8.17}$$

8.1.4 Coordination Strategy Design

Consider that a system consists of N fixed-wing UAVs governed by (8.1). An undirected graph ξ is used to describe their communication network. The cooperation control problem aims to design distributed control protocols for fixed-wing UAVs, such that groups of fixed-wing UAVs are able to collaboratively accomplish predefined flocking tasks.

The flocking algorithm is inspired from the collective behaviors of the social animals, such as fish, birds, and so on. The classical flocking algorithm consists of two parts [33]: the consensus part and the gradient-based part. The first part aims to make all UAVs move in a consistent manner. The second part mainly affects the relative positions between any two fixed-wing UAVs relying on the attractive/repulsive functions.

The classical flocking algorithm mainly faces to the first/second-order integrator model [33]. Therefore, the classical flocking algorithm cannot be directly applied to the fixed-wing UAVs because of their complex kinematical and dynamical characteristics with non-holonomic constraints. Moreover, the fixed-wing UAV model has an additional part produced by the nonzero attitude angle speed, called the gyroscopic effect. Therefore, a new distributed biological flocking algorithm is presented by adding the third part to handle the gyroscopic effect of the fixed-wing UAV. The control protocol for the fixed-wing UAV i is given by

$$\begin{aligned} U_i^F(t) &= -m_i \sum_{j \in N_i} k_{ij}\left(Q_i - Q_j\right) - m_i H_i \sum_{j \in N_i} k_{ij}\left(\dot{p}_i - \dot{p}_j\right) \\ &\quad - m_i H_i \sum_{j \in N_i(t)} \nabla_{p_{ij}} V\left(\|p_{ij}\|\right) - m_i G_i g_i \\ U_i^M(t) &= -K_i^{-1} \sum_{j \in N_i} k_{ij}\left(S_i - S_j\right) - K_i^{-1} D_i \sum_{j \in N_i} k_{ij}\left(\Theta_i - \Theta_j\right) - K_i^{-1} B_i, \end{aligned} \tag{8.18}$$

where $V\left(\|p_{ij}\|\right)$ is the potential function between UAV i and j. Therein, $-m_i G_i g_i$ is the gravity term and $-K_i^{-1} B_i$ is the gyroscopic effect item, which would disappear when the attitude angle speed vector S_i equals zero. The definition of the potential function $V\left(\|p_{ij}\|\right)$ is given as follows [26]:

Definition 8.1 (Potential Function One): Potential $V\left(\|p_{ij}\|\right)$ is a differentiable, nonnegative, radially unbounded function of the distance $\|P_{ij}\|$ between agent i and j, such that

(1) $V\left(\|P_{ij}\|\right) \to \infty$ as $\|P_{ij}\| \to 2R$;
(2) $V\left(\|P_{ij}\|\right) \to \infty$ as $P_{ij} \to D$;
(3) $V\left(\|P_{ij}\|\right)$ attains its unique minimum when agents i and j are located at a desired distance between $2R$ and D.

At this point, we present the following theorem for multiple fixed-wing UAVs.

Theorem 8.1 *Consider a system consisting of N fixed-wing UAVs with kinematics* (8.1). *The behavior of each fixed-wing UAV is steered by protocol* (8.18). *If the initial interaction network* $\xi(0)$ *is an undirected connected graph, then the following statements hold:*

1. *Any two fixed-wing UAVs cannot collide with each other; that is,* $\|P_i - P_j\| > 2R$, $i, j \in v$, and $i \neq j$.
2. *The interaction network's connectivity is preserved at all times.*
3. *The speed, the attitude angle speed, the roll angle, the pitch angle, and the yaw angle of all the fixed-wing UAVs asymptotically become the same; that is,* $u_i = u_j$, $v_i = v_j, \omega_i = \omega_j, p_i = p_j, q_i = q_j, r_i = r_j, \phi_i = \phi_j, \theta_i = \theta_j, \psi_i = \psi_j, i, j \in v, i \neq j$.
4. *The system approaches a cohesive configuration that minimizes the total potential; that is,* $\frac{dV}{dt} = 0$.

Proof According to (8.3) and (8.18), one gets

$$\begin{aligned}\dot{Q}_i &= M_i Q_i - \sum_{j\in N_i} k_{ij}(Q_i - Q_j) - H_i \sum_{j\in N_i} k_{ij}(\dot{p}_i - \dot{p}_j) - H_i \sum_{j\in N_i} \nabla_{P_{ij}} V(\|P_{ij}\|)\\ \dot{S}_i &= -\sum_{j\in N_i} k_{ij}(S_i - S_j) - D_i \sum_{j\in N_i} k_{ij}(\dot{\Theta}_i - \dot{\Theta}_j).\end{aligned} \tag{8.19}$$

Let $P = [P_1^T, ..., P_N^T]^T$, $Q = [Q_1^T, ..., Q_N^T]^T$, $\Theta = [\Theta_1^T, ..., \Theta_N^T]^T$, $S = [S_1^T, ..., S_N^T]^T$, and $m = [m_1^T, ..., m_N^T]^T$. According to the definition of the Laplacian matrix L_N, L_N is symmetric and positively semi-definite. Consider the following energy function as the common Lyapunov function:

$$\begin{aligned}E(P, \Theta, Q, S, m) &= \frac{1}{2}V + \frac{1}{2}\sum_{i=1}^{N} Q_i^T Q_i + \frac{1}{2}\sum_{i=1}^{N} S_i^T S_i + \frac{1}{2}\Theta^T (L_N \otimes I_3)\Theta\\ &= \frac{1}{2}V + \frac{1}{2}\sum_{i=1}^{N} Q_i^T Q_i + \frac{1}{2}\sum_{i=1}^{N} S_i^T S_i + \frac{1}{2}\sum_{i=1}^{N}\Theta_i^T \sum_{j\in N_i} k_{ij}(\Theta_i - \Theta_j).\end{aligned} \tag{8.20}$$

where I_3 denotes the 3×3 identity matrix. Then, the derivative of $E\left[P, \Theta, Q, S, m\right]$ w.r.t the time $t \in \left[t_r, t_{r+1}\right)$ is

$$\begin{aligned}\frac{dE}{dt} &= \frac{1}{2}\sum_{i=1}^{N}\sum_{j\in N_i} \dot{P}_i^T \nabla_{P_{ij}} V(P_{ij})\\ &+ \sum_{i=1}^{N} Q_i^T \dot{Q}_i + \sum_{i=1}^{N}\sum_{j\in N_i} \dot{\Theta}_i^T k_{ij}(\Theta_i - \Theta_j) + \sum_{i=1}^{N} S_i^T \dot{S}_i\\ &= \sum_{i=1}^{N} \dot{P}_i^T \sum_{j\in N_i} \nabla_{P_{ij}} V(\|P_{ij}\|) + \sum_{i=1}^{N} Q_i^T \dot{Q}_i\end{aligned}$$

$$+\sum_{i=1}^{N} S_i^T D_i \sum_{j\in N_i} k_{ij}(\Theta_i - \Theta_j) + \sum_{i=1}^{N} S_i^T \dot{S}_i \tag{8.21}$$

Because M_i is an anti-symmetric matrix, one gets that $\sum_{i=1}^{N} Q_i^T M_i Q_i = 0$. With (8.19) and (8.21), we have

$$\begin{aligned}\frac{dE}{dt} &= -\sum_{i=1}^{N} Q_i^T \sum_{j\in N_i} k_{ij}(Q_i - Q_j) - \sum_{i=1}^{N} S_i^T \sum_{j\in N_i} k_{ij}(S_i - S_j) - \sum_{i=1}^{N} \dot{P}_i^T \sum_{j\in N_i} k_{ij}(\dot{p}_i - \dot{p}_j) \\ &= -Q^T(L_N \otimes I_3)Q - S^T(L_N \otimes I_3)S - \mathrm{p}^T(L_N \otimes I_3)\mathrm{p}\end{aligned} \tag{8.22}$$

Hence,

$$\frac{dE}{dt} \leq 0 \tag{8.23}$$

The initial potential energy of the system is finite, and the initial speeds and initial angles of all UAVs are also finite. Thus, the initial energy $E[P(0), \Theta(0), Q(0), S(0), m(0)]$ of the system is finite. Because $\frac{dE}{dt} \leq 0$ for each time interval $[t_r, t_{r+1})$, $r = 0, 1, \ldots$, the supremum of $E(t)$ is obviously its initial value $E[P(0), \Theta(0), Q(0), S(0), m(0)]$ for all t. Then, the potential energy $V(t)$ of the system is finite. Furthermore, the potential energy $V(\|p_{ij}\|)$ between UAVs i and j is also finite.

(1) **Collision Avoidance**: Obviously, $V(\|P_{ij}\|) \to \infty$ as $\|P_{ij}\| \to 2R$, according to rule (1) of Definition 8.1. However, $V(\|P_{ij}\|) \to \infty$ violates the conclusion that $V(\|p_{ij}\|)$ remains finite. Thus, $\|P_{ij}\| > 2R$, which guarantees that any two UAVs cannot collide with each other. Conclusion 1 of Theorem 8.1 is proved.
(2) **Connectivity Preservation**: According to rule (2) of Definition 8.1, if $\|P_{ij}\| \to D$ for some $i, j \in \varepsilon$, then we have $V(\|P_{ij}\|) \to \infty$, which violates the conclusion that $V(\|p_{ij}\|)$ remains finite. Thus, one gets $V(\|P_{ij}\|) < D$. If an edge $i, j \notin \varepsilon$ is added to ε, the associated potential $V(\|P_{ij}\|)$ is bounded, and hence, so is the new potential V. Consequently, the interaction network's connectivity can be preserved. Conclusion 2 of Theorem 8.1 is proved.
(3) **Velocity Alignment**: It has been proved that the interaction network $\xi(t)$ eventually becomes fixed [14]. The following discussions are restricted on the time interval $[t_f, \infty)$, where t_f is the last switching time.

Note that the set

$$B = \{P, \Theta, Q, S, m | E(P, \Theta, Q, S, m) \leq E[P(0), \Theta(0), Q(0), S(0), m(0)]\} \tag{8.24}$$

is positively invariant. Moreover, one gets $P_{ij} < (N-1)D$ for all i and j, because $G(t)$ is connected for all $t \geq 0$. Since $E(P,\Theta,Q,S,m) \leq E\big[P(0),\Theta(0),Q(0),S(0),m(0)\big]$, one gets

$$QQ^T \leq 2E[P(0),\Theta(0),Q(0),S(0),m(0) \tag{8.25}$$

that is, $\|Q\| \leq \sqrt{2E\big[P(0),\Theta(0),Q(0),S(0),m(0)\big]}$。另外，$\Phi_i \in (-\pi,\pi)$, $\theta_i \in (-\pi,\pi)$, $\Psi_i \in (-\pi,\pi)$, $i=1,\ldots,N$. In addition, $\phi_i \in (-\pi,\pi)$, $\theta_i \in \left(-\frac{\pi}{2},\frac{\pi}{2}\right)$, $\psi_i \in (-\pi,\pi)$, $i=1,\cdots,N$. Thus, the set B is closed and bounded, hence compact. As an autonomous system on the connected time interval $\left[t_f,\infty\right)$, the system (8.1) with control input (8.18) will converge to the invariant set $IS = \left\{P,\Theta,Q,S,m\middle|\frac{dE}{dt}=0\right\}$, according to the LaSalle-Krasovskii invariance principle [34]. Clearly, $\frac{dE}{dt}=0$ if and only if $Q_i=Q_j$, $P_i=P_j$, and $\dot{P}_i=\dot{P}_j$ for $i=1,\ldots,N$ and $j \in N_i$.

According to (8.1) and $Q_i=Q_j$, $\dot{P}_i=\dot{P}_j$ is equivalent to $H_i=H_j$, for $i,j=1,\ldots,N$. Then, we obtain $\Theta_i=\Theta_j$.

With $k_{ij}=k_{ji}$, one gets $\sum_{i=1}^{N}\sum_{j\in N_i}k_{ij}\big(\Theta_i-\Theta_j\big)=0$. From (8.19) as well as $S_i=S_j$, one has

$$\begin{aligned}\dot{\overline{S}}_i &= \frac{1}{N}\sum_{i=1}^{N}\dot{S}_i\\ &= \frac{1}{N}\left(-\sum_{i=1}^{N}\sum_{j\in N_i}k_{ij}(S_i-S_j)-\sum_{i=1}^{N}\sum_{j\in N_i}k_{ij}\big(\Theta_i-\Theta_j\big)\right)\\ &= -\frac{1}{N}\sum_{i=1}^{N}\sum_{j\in N_i}k_{ij}\big(\Theta_i-\Theta_j\big)=0.\end{aligned} \tag{8.26}$$

Because $S_1=\ldots=S_N=\overline{S}$, we have $\overline{S}_i=\dot{\overline{S}}=0$; that is, $\sum\limits_{j\in N_i}k_{ij}\big(S_i-S_j\big)-\sum\limits_{j\in N_i}k_{ij}\big(\Theta_i-\Theta_j\big)=0$, $i=1,\ldots,N$. With $S_i=S_j$ and $k_{ij}=k_{ji}>0$, one gets $\Theta_i=\Theta_j$ for $i=1,\ldots,N$. Thus, conclusion 3 of Theorem 8.1 is proved.

(4) **Flocking Center:** $\nabla_{P_{ij}}V\big(\|P_{ij}\|\big)$ is an odd function. $Q_i=Q_j$ and $\dot{P}_i=\dot{P}_j$. From (8.19), one has

$$\begin{aligned}\dot{\overline{Q}} &= \frac{1}{N}\sum_{i=1}^{N}\dot{Q}_i\\ &= \frac{1}{N}\left(-\sum_{i=1}^{N}\sum_{j\in N_i}k_{ij}(Q_i-Q_j)-\sum_{i=1}^{N}H_i\sum_{j\in N_i}k_{ij}(\dot{p}_i-\dot{p}_j)-\sum_{i=1}^{N}H_i\sum_{j\in N_i}\nabla_{P_{ij}}V\big(\|P_{ij}\|\big)\right)\end{aligned}$$

$$= \frac{1}{N}\left(-\sum_{i=1}^{N} H_i \sum_{j\in N_i} \nabla_{P_{ij}} V\left(\|P_{ij}\|\right)\right)$$
$$= 0. \tag{8.27}$$

Because $Q_1 = \ldots = Q_N = \overline{Q}$, we have $\overline{Q}_i = \dot{\overline{Q}} = 0,$; that is, $-\sum_{j\in N_i} k_{ij}(Q_i - Q_j) - H_i \sum_{j\in N_i} k_{ij}\left(\|\dot{P}_i - \dot{P}_j\|\right) - H_i \sum_{j\in N_i} \nabla_{P_{ij}} V \|P_{ij}\| = 0$, $i = 1, \ldots, N$. With $Q_i = Q_j$ and $\dot{P}_i = \dot{P}_j$, one gets $\nabla_{P_{ij}} V\left(\|P_{ij}\|\right) = 0$ for $i = 1, \ldots, N$, $j \in N_i$. Then, because the total potential of the UAV system is $V(t) = \sum_{i=1}^{N} \sum_{j\in N_i} V\left(\|P_{ij}\|\right)$, one gets $\frac{dV}{dt} = 0$. Conclusion 4 of Theorem 8.1 is proved.

8.1.5 Simulation Validation

In this section, we discuss the steady flight scene of fixed-wing UAVs with horizontal aerofoils, whose additional constraint conditions are given by

$$\phi_i = \dot{\phi}_i = \dot{\theta}_i = \dot{\psi}_i, i = 1, \ldots, N \tag{8.28}$$

Thus, when the fixed-wing UAV system reaches a steady state, $\phi_i = p_i = q_i = r_i \equiv 0$ is expected. Then, the simplified control protocol for the fixed-wing UAV i is represented as

$$\begin{aligned}
U_i^F(t) &= -m_i \sum_{j\in N_i} k_{ij}(Q_i - Q_j) - m_i H_i \sum_{j\in N_i} k_{ij}(p_i - \dot{p}_j) \\
&\quad - m_i H_i \sum_{j\in N_i(t)} \nabla_{p_{ij}} V\left(\|p_{ij}\|\right) - m_i G g_i \\
U_i^M(t) &= -K_i^{-1} \sum_{j\in N_i} k_{ij}(S_i - S_d) - K_i^{-1} \sum_{j\in N_i} k_{ij}(S_i - S_j) - K_i^{-1} \sum_{j\in N_i} k_{ij}(_d) \\
&\quad - K_i^{-1} \sum_{j\in N_i} k_{ij}(\Theta_i - \Theta_j) - K_i^{-1} B_i.
\end{aligned} \tag{8.29}$$

where the desired angle speed vector $S_d = [p_d, q_d, r_d]^T = 0$ and $\Theta_d = [\phi_d, \theta_d, \psi_j]^T$. The desired roll angle speed $\phi_d = 0$.

For the sake of verifying the effectiveness of the proposed algorithm (8.29), we employ six fixed-wing UAVs governed by (8.1). The initial conditions of the six UAVs are generated randomly but are content with the condition that the initial communication network is connected. Given that the communication radius $D = 10m$, based on the initial positions of six UAVs and the communication radius, the initial interaction topology of the UAV system is obtained as

$$A(0) = \begin{bmatrix} 0 & 0 & 1 & 1 & 1 & 1 \\ 0 & 0 & 0 & 0 & 1 & 1 \\ 1 & 0 & 0 & 1 & 0 & 0 \\ 1 & 0 & 1 & 0 & 0 & 0 \\ 1 & 1 & 0 & 0 & 0 & 1 \\ 1 & 1 & 0 & 0 & 1 & 0 \end{bmatrix} \tag{8.30}$$

which is connected, but not completed.

The potential function consists of two parts: attractive potential and repulsive potential. $a \in R^+$ denotes the coefficient of the attractive potential, whereas $b \in R^+$ expresses the coefficient of the repulsive potential.

Each UAV's physical diameter is $2R = 9.45$ m. The repulsive potential coefficient is $b = 1$. The simulation time is $tf = 500$ s. The evaluation of the coupling strength k_{ij} is assigned by (8.16), where $k = C_k \frac{D}{\sqrt{a}}$. Here, C_k is a positive constant, and $C_k = 1.5$.

The attractive potential coefficient a is one of the factors who can decided the desired balance distance d_{dir} [35]. Meanwhile, a influences the convergence speed of the whole system. If a has a small value, then the whole system converges to a consensus state at a slow speed. In order to shorten the convergence time, a is expected to have a large value. But this large value may cause the final balance distance do not satisfy our requirement. Therefore, a is expressed by a linear time-variant function. According to this time-variant function, a has an initial value at the beginning, and then becomes smaller and smaller following with time. Finally, a reaches a proper value which is decided by the desired balance distance. Then, $a(t)$ can be expressed by

$$a(t) = 2500 - (2500 - d_{sir})\frac{t}{tf} \tag{8.31}$$

Since $d_{sir} \ll 2500$, (8.31) can be approximately written as

$$a(t) = 2500 - 5t \tag{8.32}$$

Then, the simulational results of six fixed-wing UAVs governed by (8.1) under the control protocol (8.29) are shown in Fig. 8.1a, b.

First, Fig. 8.1a shows the trajectories of the six UAVs under the proposed algorithm. The green star point denotes the initial position of each UAV, and the colorful ball is the final position of each UAV at time $t = 500$ s. Figure 8.1a also gives a clearer exhibition of the final configuration of the six UAVs at time $t = 500$ s, which is a three-dimensional cohesive formation. Furthermore, Fig. 8.1b gives the distances between any two UAVs. Because these distances converge to stable, this also further proves that the system configuration becomes stable. Besides, the minimum distance between any two UAVs is larger than the safe distance $2R = 9.45$ m; thus, no collision occurs between any two UAVs. Third, Fig. 8.2 give the variances of the roll

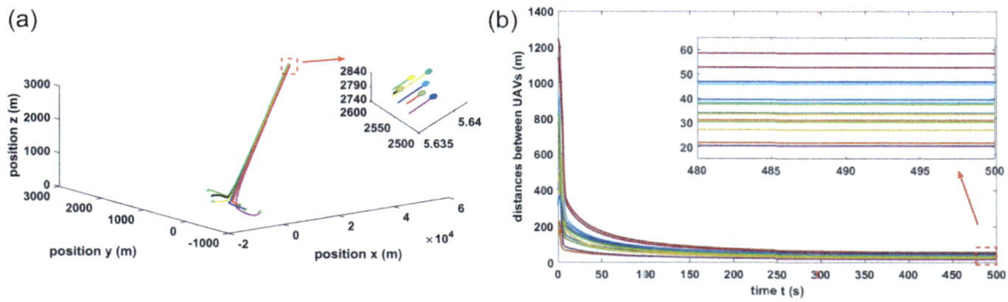

Fig. 8.1 The trajectories of six fixed-wing UAVs for steady fight flocking at variable altitudes. **a** trajectories of six UAVs **b** distances between any two UAVs

angle, pitch angle, yaw angle, x-axis speed component, y-axis speed component, and z-axis speed component of each fixed-wing UAV versus time. These figures tell us that the attitude angles and the speeds of each fixed-wing UAV converge with those of the others.

More often, we expect that these fixed-wing UAVs to form a cohesive flocking at a fixed altitude. Thus, we can further improve the simplified control protocol (8.29) for the fixed-wing UAV i as

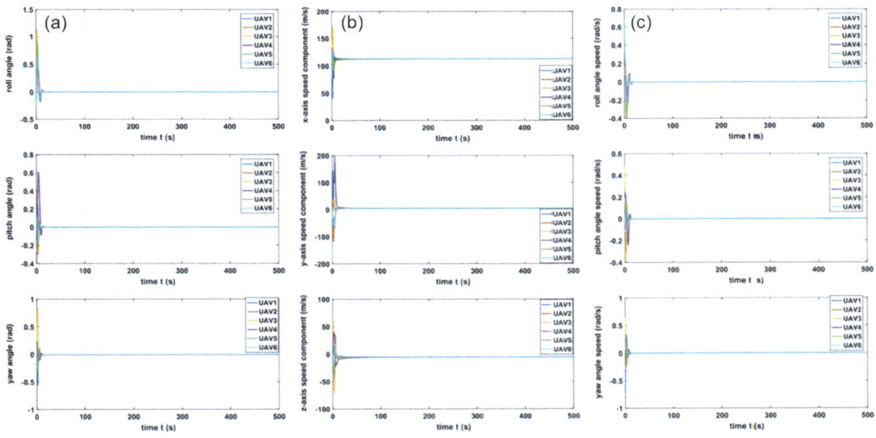

Fig. 8.2 Simulation results on state variables of six fixed-wing UAVs for steady flight flocking at variable altitudes. **a** attitude angle **b** speed **c** attitude angle speed

$$
\begin{aligned}
U_i^F(t) = &-m_i \sum_{j \in N_i} k_{ij}(Q_i - Q_d) - m_i \sum_{j \in N_i} k_{ij}(Q_i - Q_j) \\
&- m_i H_i \sum_{j \in N_i} k_{ij}\left(\dot{P}_i - \dot{P}_j\right) - m_i H_i \sum_{j \in N_i(t)} \nabla_{p_{ij}} V\left(\|p_{ij}\|\right) - m_i G g_i \\
U_i^M(t) = &-K_i^{-1} \sum_{j \in N_i} k_{ij}(S_i - S_d) - K_i^{-1} \sum_{j \in N_i} k_{ij}(S_i - S_j) \\
&- K_i^{-1} \sum_{j \in N_i} k_{ij}(\Theta_i - \Theta_j) - K_i^{-1} B_i,
\end{aligned}
\tag{8.33}
$$

where the desired speed vector $Q_d = [u_i, v_i, \omega_d]^T$. Therein, $\omega_d = 0$. The simulation results of the trajectories of six fixed-wing UAVs (8.1) under the control protocol (8.33) are shown in Fig. 8.3a. Figure 8.3b gives the distance between any two UAVs. Figure 8.3c and d verify that the flight altitude of the six fixed-wing UAVs is fixed during the steady flight.

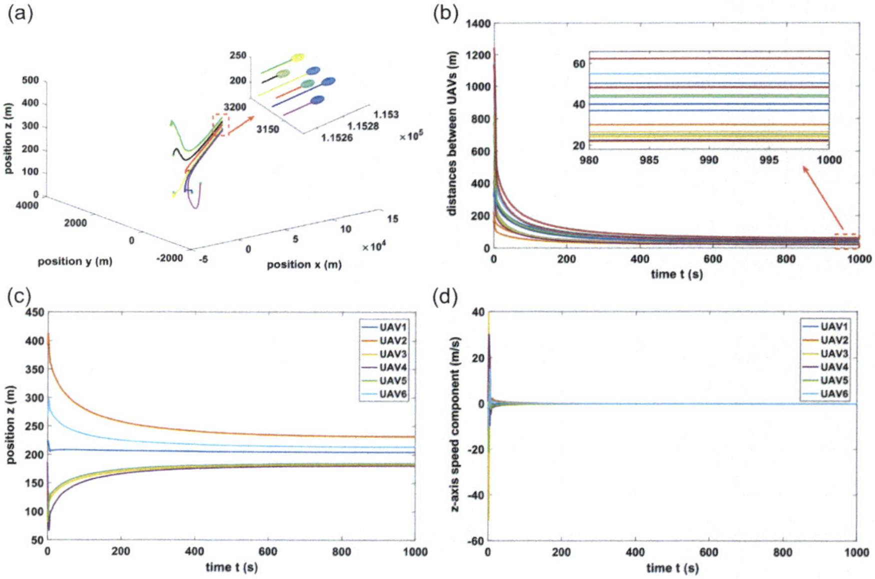

Fig. 8.3 The trajectories of six fixed-wing UAVs for steady flight flocking at invariable altitude. **a** trajectories of six UAVs **b** distances between any two UAVs **c** altitude versus time **d** z-axis speed component versus time

8.1.6 Discussion and Analysis

In this part, we focus on discussing C_k, the coefficient of the coupling strength. First, we explain how to find the value of C_k. Second, we discuss the influence of the value of C_k on the evolution process of the coordination system.

The first criterion for deciding the value of C_k is that it should ensure that any two UAVs avoid collision with each other. Let $P_{\min}$ denote the minimum distance between any two UAVs during the evolution process. Then, $P_{\min}$ just needs to be larger than the safe distance $2R = 9.45$ m to avoid collision. However, the airflow disturbance exists among neighbor UAVs. Therefore, to avoid the influence coming from the airflow disturbance on the behaviors of the UAVs, we consider that the following constraint (8.34) to be satisfied:

$$P_{min} \leq 20\,\text{m} \tag{8.34}$$

To find the proper value of C_k to satisfy the constraint (8.34), we give 20 sample values to C_k from 0.5 to 10. Then, we record the value of $P_{\min}$ during each system evolution process (8.1) under the control protocol (8.29) for each sample value of C_k. The results are shown in Fig. 8.4a. Clearly, $P_{\min}$ grows along with the increase of C_k from 0.5 to 10. When $C_k \geq 1.5$, $P_{\min} \geq 50$. These data tell us one preliminary conclusion: the bigger C_k may produce a bigger $P_{\min}$ during a certain interval. This conclusion is very helpful for adjusting the coordination system to allow for safe flight.

C_k is definitely also expected to force the system to reach a stable state as soon as possible. Thus, the adjustment time t^* becomes the second criterion for selecting the parameter C_k. Adjustment time t^* is commonly used to depict the time at which one system reaches stable from initial conditions, and it is also an important indicator for evaluating the control effect of the cooperation algorithm. Let the error band $e_{ss} = 0.1$ m. Then, the adjustment time t^* is the smallest value of T that satisfies (8.35)

$$\left\{\|P_{ij}(t) - P_{ij}(tf\,\|) \leq e_{ss}\right\} \tag{8.35}$$

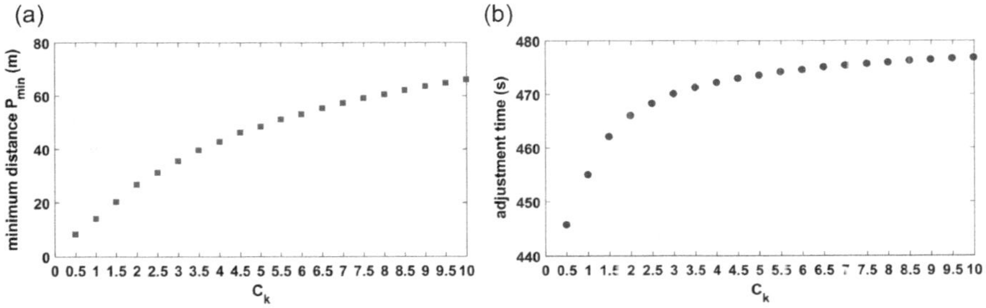

Fig. 8.4 Analysis results. **a** C_k versus $P_{\min}$ **b** C_k versus t^*

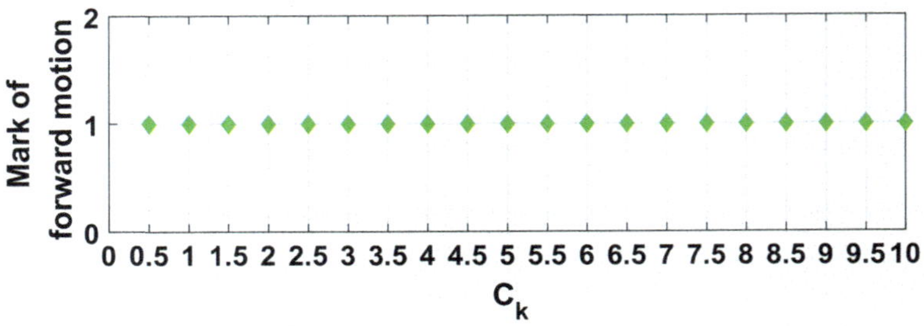

Fig. 8.5 C_k versus mark of forward motion

for all $i, j \in \varepsilon$, $T \leq t \leq t_f$. Then, the smaller the adjustment time t^* is, the faster the system reaches steady state and the better the control effect of the cooperation algorithm.

Similarly, we also record the value of t^* during every system evolution process (8.1) under the control protocol (8.29) for each sample value of C_k from 0.5 to 10. The results are shown in Fig. 8.4b. Clearly, t^* grows with the increase of C_k from 0.5 to 10. These data tell us another preliminary conclusion: a smaller C_k may cause a bigger rate of convergence during a certain interval. This conclusion gives a totally opposite expectation for the value of C_k.

According to Fig. 8.5, the marks of forward motion are all equal to 1, which means that a forward speed of each fixed-wing UAV can be obtained for each flight under different values of C_k. It seems that the forward speed of each UAV depends on the initial station of the six fixed-wing UAVs, but has nothing to do with the selection of the value of C_k.

We also record the maximal value of the speed vector Q_i (see Fig. 8.6), the maximal value of the attitude angle vector Θ_i (see Fig. 8.7), the maximal value of the attitude angle speed vector S_i (see Fig. 8.8) under different values of C_k from 0.5 to 10, for $i = 1, \ldots, 6$. Obviously, the maximal value decreases with the increase of the value of C_k. Thus, the bigger value of C_k is preferred for those fixed-wing UAVs, which have smaller upper bounds of speed, attitude angle, or attitude angle speed.

Above all, we can choose a proper value for C_k according to the specific constraint of the chosen fixed-wing UAVs from the aspects of adjustment time, minimal inter-agent distance $P_{\min}$, and the bounds of Θ_i, Q_i, and S_i $(i = 1, \cdots, N)$. To satisfy the constraints (8.34), we choose $C_k = 1.5$ for variable-altitude flocking, while we choose $C_k = 2.5$ for invariable-altitude flocking.

8.1.7 Summary

This section solved the fixed-wing UAV flocking problem of the fixed-wing UAV governed by the complete six-degrees-of-freedom model. The weighted communication topology of the UAV system is considered to be time-variant according to

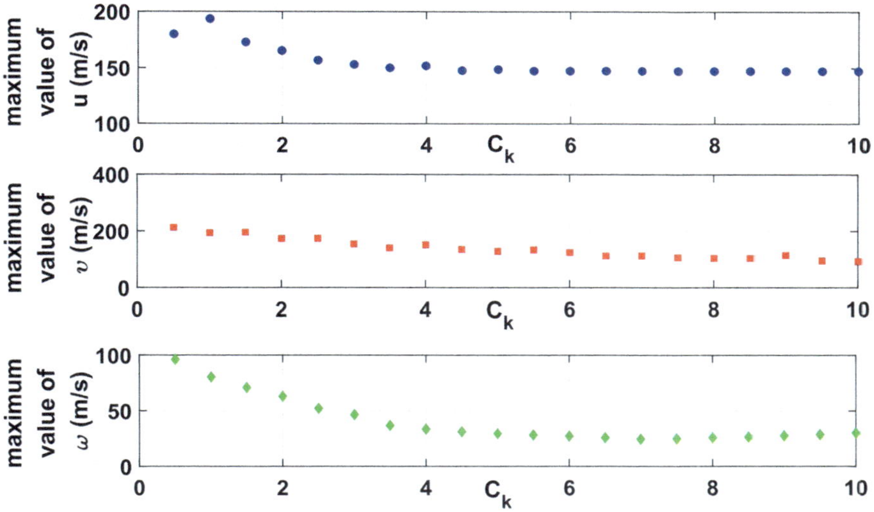

Fig. 8.6 C_k versus maximum value of speed

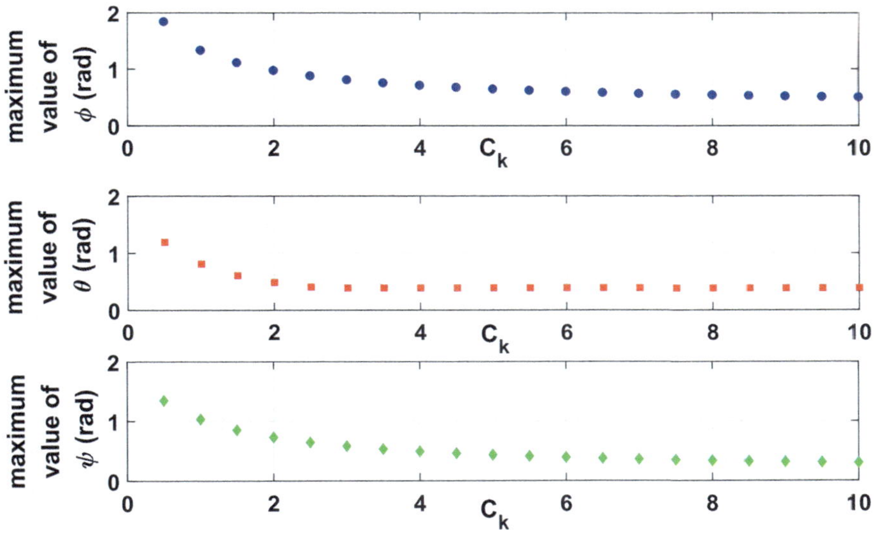

Fig. 8.7 C_k versus maximum value of attitude angle

the nearest neighbor rules among individuals. Based on the proposed model and the communication mechanism, a distributed cooperation approach is designed to force groups of fixed-wing UAVs to collaboratively accomplish such predefined flocking tasks like flocks of birds. As defined by Reynolds, 3 essential attributes, separation, alignment, and cohesion, are all proved to be satisfied. Furthermore, 4 constraint

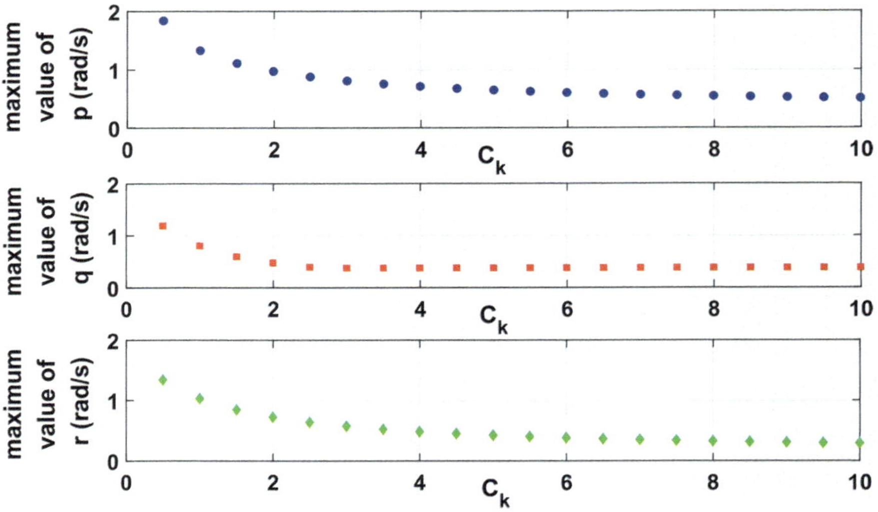

Fig. 8.8 C_k versus maximum value of attitude angle speed

conditions are considered in this system, including bound of attitude angle, velocity, and attitude angle speed; minimal inter-agent distance; forward motion; and adjustment time. These constraints are tactfully considered as evaluation criteria to designate the value of the communication network coupling strength. Finally, 2 kinds of simulations are respectively implemented on six fixed-wing UAVs to validate the effectiveness of the proposed algorithm for the steady flight task of variable-altitude flocking and the steady flight task of invariable-altitude flocking. Moreover, this distributed system is easily scalable, from several UAVs to thousands of UAVs.

8.2 Optimization of Communication Radius

Herein, we present a methodology and framework for exploiting certain interdisciplinary studies that can particularly benefit from integration. In this paper, rigorous derivation of control theory and statistical analysis of simulation results are organically unified for testifying and optimizing the emergence of order in aerial swarming scenarios under free boundary conditions. Each Unmanned Aerial Vehicle (UAV) is regulated by a simplified mathematical model, based on which a distributed flocking protocol is proposed as a feasible solution for aerial swarms. On condition that the initial interaction network is connected, the LaSalle-Krasovskii invariance principle is implemented to verify the effectiveness of the above algorithm.

However, most existing results on flocking are far from being engineering applications. A basic challenge is how to present a low-cost energy and time saving solution on account of the limited flight capability of these UAVs and real-time operational requirements. As is well known, energy consumption can be reduced if unnecessary interactions among individuals are eliminated. Therefore, another contribution of this paper is to propose a precise optimization of an existing flocking algorithm for UAVs with respect to interaction requirements. Energy and time measurements, as well as scalability effects, are assessed in terms of statistical significance and strength. The results indicate that the flocking control protocol adopting the minimal interaction is the most promising swarm.

8.2.1 Introduction

Aerial swarm has been considered as one of the most challenging, exciting, and multidisciplinary fields of robotics in the last decades [36–44]. Compared with single-robot systems, aerial swarm systems are expected to be more robust to failure and own faster response capability to complete complex tasks, such as patrolling, exploration, and search and rescue in large areas. Flocking is a type of collective behavior emerging from a set of individuals that rely only on simple rules and local sensing. In recent years, flocking has been widely studied as an effective and low-cost solution to aerial swarm. However, the computation complexity and quantity will increase sharply following the expansion of the scale of the group. Therefore, it is necessary to discuss how to accelerate the emergence of flocking structures.

The literature on flocking can be summarized into two categories. One category is to discuss the flocking problem by using rigorous mathematical analysis [45–49]. There are mainly three basic elements to study the flocking behavior in this category, including model, interaction rules, and distributed control protocol. Prior studies on flocking are mainly based on either the simple first-order or second-order integrator model or a general dynamic model [50–52]. The nonlinear dynamics inherent in the system of Unmanned Aerial Vehicles (UAVs) remain sparse, which can, with a large probability, cause the failure of the proposed control algorithm to resolve the coordination control of UAVs. A kind of scale-dependent interaction rule is usually adopted, called the nearest-neighbor interaction rule [53]. The mechanism of the interaction network depends on communication radius. Inspired by natural starlings, a flocking algorithm with the combination of a consensus and artificial potential field has been proposed to reveal the underlying mechanism responsible for the emerging collective behaviors that a set of individuals move in unison while creating a dynamic formation pattern. The flocking control of wheeled mobile robots [54] and underwater robots [49, 55] were dealt with, yet these studies just considered the two-dimensional motion of individuals. Jia et al. have studied the three-dimensional flocking of UAVs governed by a nonlinear dynamic model, but did not give any further analysis and discussion on how to accelerate the emergence of flocking behaviors and how to apply the proposed control algorithm to large-scale systems [56].

The second category uses a statistical method to analyze the simulation results. These simulations are designed mostly according to the real experimental data of flocks of animals. Ballerini et al. have tracked and analyzed the movements of these starling flocks [57]. In this study, they found that there is not any leader in the colony. Besides, the number of individual birds in a flock can be as many as thousands, although each bird just follows local rules and only interacts with six to seven neighbors. Generally speaking, bigger interaction networks (referring to more interactions) yield faster consensuses. However, starlings show fast and flexible dynamic aggregation relying only on small interaction networks. Therefore, a problem exists in how the scale of the interaction network influences the emergence behavior of a flock. As proposed by economists, Rybski et al. found a phenomenon called diminishing returns, which led by the increasing of the number of deployed robots [58]. Rosenfeld et al. also found that additional robots usually produce negative returns after a peak in performance when they focused on the scalability study of foraging tasks [59]. Besides, Timothy et al. have compared three strategies for aerial swarm deployment task according to the energy-time efficiency by using statistical analysis methods [60]. In this paper, a statistical method is applied to analyze whether the diminishing returns exist in the interaction network of an aerial swarm or not.

According to the above research results, rigorous theoretical analysis can offer rigorous mathematical proof, but these conclusions are always limited to some unreasonable hypothesis. At the same time, although simulation can be carried out as close as possible to realistic conditions, these conclusions are easily doubted and argued because of some special cases. Therefore, our goal here is to design a general methodology and framework for studying flocking behaviors by means of the combination of control theory and statistical method.

First, the UAV system is considered to be an egalitarian one with two basic properties: One is that every individual obeys the same control protocol to adjust its behavior during the entire process of flight, and the other is that each UAV uses the same communication radius to recognize its neighbors for information exchange. Thus, the communication radius can be regarded as the only factor to draw the scale of the interaction network. Second, a flocking protocol is proposed for each UAV to form a three-dimensional cohesive configuration in a consistent pattern. Instead of a simple first-order or second-order integrator, each UAV is simplified as a nonlinear dynamic model commonly used to describe the UAV in the existing literature [61, 62]. According to the LaSalle-Krasovskii invariance principle [63], the UAV flocking will emerge as long as the initial interaction network is connected. During the evolutionary process, collision is carefully avoided among pairwise UAVs. Finally, energy/time costs are evaluated under different communication radii, and the scalability performance is examined by increasing the group size. A statistical method is applied to analyze the energy/time measurements and scalability effects. The results indicate that the flocking protocol adopting the minimally connected communication radius is considered as the most promising swarm from the perspective of energy savings and the rate of convergence. This conclusion tactfully echoes from the biological evidence that each starling interacts with only six or seven nearest-neighbors most often at the same time.

8.2.2 *Materials and Methods*

8.2.2.1 UAV Model

A collective task is considered to be carried out by N homogeneous UAVs flying in three-dimensional space. Each UAV has the same hardware installment and software configuration, as well as owns the same communication capability. The communication capability of each UAV can be measured by its communication radius D (D is a positive constant). According to the existing literatures on UAVs, the dynamics of each UAV is described by the following differential equation [61, 62]

$$\begin{aligned}
\dot{x}_i(t) &= \vartheta_i(t)\cos\gamma_i(t)\cos\psi_i(t) \\
\dot{y}_i(t) &= \vartheta_i(t)\cos\gamma_i(t)\sin\psi_i(t) \\
\dot{h}_i(t) &= \vartheta_i(t)\sin\gamma_i(t) \\
\dot{\theta}_i(t) &= g\cdot(n_{ti}(t)-\sin\gamma_i(t)) \\
\dot{\psi}_i(t) &= g\cdot n_{hi}(t)/(\vartheta_i(t)\cos\gamma_i(t)) \\
\dot{\gamma}_i(t) &= g\cdot(n_{vi}(t)-\cos\gamma_i(t))/\vartheta_i(t),\ i=1,\ldots,N
\end{aligned} \tag{8.36}$$

where i denotes index for the ith UAV, $\dot{x}_i(t)$ is the ith UAV's downrange, $\dot{y}_i(t)$ is the ith UAV's crossrange, $h_i(t)$ is the ith UAV's altitude, $\vartheta_i(t) > 0$ is the ith UAV's speed, $\psi_i(t) \in \left(-\frac{\pi}{2}, \frac{\pi}{2}\right)$ is the ith UAV's heading angle, $\gamma_i(t) \in \left(-\frac{\pi}{2}, \frac{\pi}{2}\right)$ is the ith UAV's flight path angle, $n_{hi}(t)$ is the ith UAV's horizontal load factor, $n_{vi}(t)$ is the ith UAV's vertical load factor, $n_{ti}(t)$ is the ith UAV's tangential load factor, and g is the constant acceleration of gravity.

For brevity, the proposed simplified model (8.36) of UAV i is shown in Fig. 8.9. Generally speaking, the behavior of UAV i is steered under four forces: lift force, drag force, weight, and thrust force. $T_i(t)$ is the magnitude of the thrust, $L_i(t)$ is the magnitude of the lift force, $D_i(t)$ is the magnitude of the drag force, and $W_i(t)$ is the magnitude of the gravity. Heading angle $\psi_i(t)$ describes whether the aircraft is turning left or right. Flight path angle $\gamma_i(t)$ describes whether the aircraft is climbing or descending. Bank angle $\mu_i(t) = arctan\frac{n_{hi}(t)}{n_{vi}(t)}$ describes whether the aircraft is rolling clockwise or counterclockwise. The three-dimensional aircraft model not only considers the longitudinal plane, but also the lateral plane, thus the control variable $[L_i(t), D_i(t), T_i(t)]$ is replaced with $[n_{ti}(t), n_{hi}(t), n_{vi}(t)]$ in this model (8.36).

Then, the state of UAV i at time t can be expressed by its position vector $P_i(t) = \left[x_i(t), y_i(t), h_i(t)\right]^T$ and its velocity vector

$$q_i(t) = \begin{bmatrix} \vartheta_i(t)\cos\gamma_i(t)\cos\psi_i(t) \\ \vartheta_i(t)\cos\gamma_i(t)\sin\psi_i(t) \\ \vartheta_i(t)\sin\gamma_i(t) \end{bmatrix} \tag{8.37}$$

Thus, we have

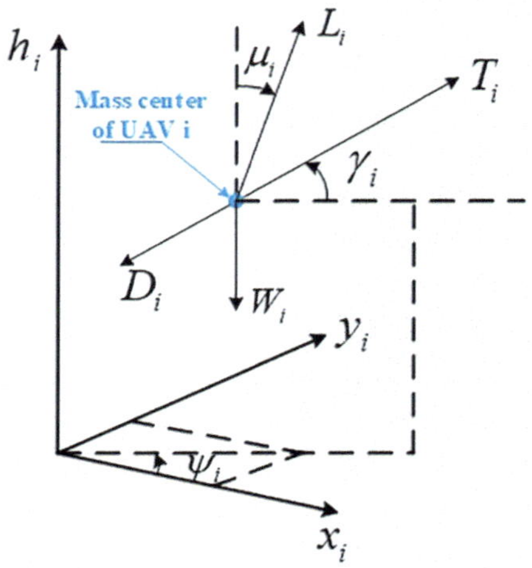

Fig. 8.9 The simplified dynamical model of Unmanned Aerial Vehicle (UAV) i[62]

$$\dot{q}_i(t) = \begin{bmatrix} \dot{\vartheta}_i \cos\gamma_i \cos\psi_i - \dot{\gamma}_i\vartheta_i \sin\gamma_i \cos\psi_i - \dot{\psi}_i\vartheta_i \cos\gamma_i \sin\psi_i \\ \dot{\vartheta}_i \cos\gamma_i \sin\psi_i - \dot{\gamma}_i\vartheta_i \sin\gamma_i \sin\psi_i + \dot{\psi}_i\vartheta_i \cos\gamma_i \cos\psi_i \\ \dot{\vartheta}_i \sin\gamma_i + \dot{\gamma}_i\vartheta_i \sin\gamma_i \end{bmatrix} \tag{8.38}$$

Here, due to the space limitation, we use $q_i, \theta_i, \gamma_i, \psi_i$ instead of $q_i(t), \theta_i(t), \gamma_i(t), \psi_i(t)$ for short. Put the last three equations of (8.36) into (8.38), one gets

$$\dot{q}_i = g \begin{bmatrix} n_{ti} \cos\gamma_i \cos\psi_i - n_{hi} \sin\psi_i - n_{vi} \sin\gamma_i \cos\psi_i \\ n_{ti} \cos\gamma_i \sin\psi_i + n_{hi} \sin\psi_i \\ n_{ti} \sin\gamma_i + n_{vi} \cos\gamma_i - 1 \end{bmatrix} \tag{8.39}$$

Let the control input of UAV i at time t is denoted as $u_i(t) = [n_{ti}(t), n_{hi}(t), n_{vi}(t)]^T$. Then, Eq. (8.39) can be rewritten as

$$\dot{q}_i = g \begin{bmatrix} \cos\gamma_i \cos\psi_i & -\sin\psi_i & -\sin\gamma_i \cos\psi_i \\ \cos\gamma_i \sin\psi_i & \cos\psi_i & -\sin\gamma_i \sin\psi_i \\ \sin\gamma_i & 0 & \cos\gamma_i \end{bmatrix} \begin{bmatrix} n_{ti}(t) \\ n_{hi}(t) \\ n_{vi}(t) \end{bmatrix} - G_r \tag{8.40}$$

where $G = [0, 0, g]^T$. Then, the simplified point-mass model of UAV i can be rewritten by

$$\begin{aligned} \dot{p}_i(t) &= q_i(t) \\ q_i(t) &= gH_i^T(t)u_i(t) - G_r i = 1, \ldots, N \end{aligned} \tag{8.41}$$

where $H_i(t) = [t_i(t), n_i(t), h_i(t)]^T$. The definitions of $t_i(t), n_i(t), h_i(t)$ are, respectively,

$$t_i(t) = \begin{bmatrix} \cos\gamma_i \cos\psi_i \\ \cos\gamma_i \sin\psi_i \\ \sin\gamma_i \end{bmatrix} \tag{8.42}$$

$$n_i(t) = \begin{bmatrix} -\sin\psi_i \\ \cos\psi_i \\ 0 \end{bmatrix} \tag{8.43}$$

$$h_i(t) = \begin{bmatrix} -\sin\gamma_i \cos\psi_i \\ -\sin\gamma_i \sin\psi_i \\ \cos\gamma_i \end{bmatrix} \tag{8.44}$$

Obviously, $H_i(t)$ is an orthogonal matrix. $t_i(t), n_i(t), h_i(t)$ are orthogonal unit vectors.

8.2.2.2 Interaction Mechanism

As mentioned above, the UAV system involved in this paper is supposed to be an egalitarian system. For egalitarian systems, the interaction between any two individuals is undirectional. Thus, the metric interaction network $\xi(t)$ of the UAV system is definitely a dynamic undirected graph consisting of a vertex set $v = \{1, \ldots, N\}$ and a time-varying edge set $\varepsilon(t) = \{(i,j)|(i,j) \in v \times v, j \in N_i(t)\}$. Therein, each vertex denotes a UAV, while each edge denotes the pairwise–interaction relationship of two UAVs. $N_i(t)$ is defined as the neighbor set of UAV i at time t, $i = 1, \ldots, N$. The initial value of $N_i(t)$ is described by the following formula,

$$N_i(0) = \{j | \|p_{ij}(0)\| < D, j = 1, \ldots, N, j \neq i\}, \tag{8.45}$$

where $p_{ij}(t) = p_i(t) - p_j(t)$, $\|\cdot\|$ denotes the Euclidean norm. That is to say, each UAV is assumed to interact with its neighbor UAVs.

The interaction network $\xi(t)$ is supposed to switch at t_p, $p = 1, 2, \cdots$. Thus, $\xi(t)$ is fixed in each time-interval $[t_r, t_{r+1})$, where $r = 0, 1, \cdots$. $\xi(0)$ is assumed to be connected. During the whole evolution process, the switch condition satisfying the following hysteresis [64]

(1) When $(i,j) \notin \varepsilon(t^-)$, if $\|p_{ij}(t)\| < D - \zeta$, $\zeta \in (0, D)$, then $(i,j) \in \varepsilon(t)$, for $t > 0$;
(2) When $(i,j) \notin \varepsilon(t^-)$, if $\|p_{ij}(t)\| < D$, then $(i,j) \in \varepsilon(t)$, for $t > 0$.

8.2.2.3 Control Protocol and Stability Analysis

In 1987, Reynolds developed a computer model mimicking animal aggregation and social cohesion in animal groups [30]. This model consists of three steering behaviors--collision avoidance, velocity alignment, and ~ flocking cohesion, which describe how an individual agent maneuvers based on the relative positions and relative velocities of its nearby flock mates. Following this work, several variants on flocking model has been proposed [46, 49, 53, 65–67], and led to derive decentralized controllers consisting of consensus algorithm and artificial potential field which provably give rise to these collective phenomena. According to the specific UAV model (8.36), the control protocol of each UAV i, $i = 1, \ldots, N$ is proposed as

$$\begin{aligned}
n_{ti}(t) &= -\frac{1}{g} t_i^T(t) \sum_{j \in N_i(t)} \left(q_i(t) - q_j(t)\right) + \frac{1}{g} t_i^T(t) G - \frac{1}{g} t_i^T(t) \sum_{j \in N_i(t)} \nabla_{p_{ij}(t)} V\left(\|p_{ij}(t)\|\right) \\
n_{hi}(t) &= -\frac{1}{g} n_i^T(t) \sum_{j \in N_i(t)} \left(q_i(t) - q_j(t)\right) + \frac{1}{g} n_i^T(t) G - \frac{1}{g} n_i^T(t) \sum_{j \in N_i(t)} \nabla_{p_{ij}(t)} V\left(\|p_{ij}(t)\|\right) \\
n_{vi}(t) &= -\frac{1}{g} h_i^T(t) \sum_{j \in N_i(t)} \left(q_i(t) - q_j(t)\right) + \frac{1}{g} h_i^T(t) G - \frac{1}{g} h_i^T(t) \sum_{j \in N_i(t)} \nabla_{p_{ij}(t)} V\left(\|p_{ij}(t)\|\right)
\end{aligned} \tag{8.46}$$

There are three equations in control protocol (8.46). The first part of each equation is the consensus term, which aims at forcing all the UAVs fly with consistent velocities. The second part of each equation is the gravity term. The third part is the negative gradient of the potential function $V\left(\|p_{ij}(t)\|\right)$, which is used to adjust the relative positions between any two UAVs. The potential function can be divided into a repulsive item and an attractive item, whose definition is given as follows.

Definition 8.2 (Potential function two): Potential $V\left(\|p_{ij}(t)\|\right)$ is a differentiable, non-negative, radially unbounded function of the distance $p_{ij}(t)$ between agents i and j, such that

(1) $V\left(\|p_{ij}(t)\|\right) \to \infty$ as $\|p_{ij}(t)\| \to 0$;
(2) $V\left(\|p_{ij}(t)\|\right) \to \infty$ as $\|p_{ij}(t)\| \to D$;
(3) $V\left(\|p_{ij}(t)\|\right)$ attains its unique minimum when agents i and j are located at a desired distance between 0 and D.

Rule (1) of Definition 8.2 is related to the repulsive item, which forces the UAVs to avoid collision with each other. Rule (2) of Definition 8.2 is related to the attractive item, which mainly affects the flocking cohesion behavior of these UAVs. Rule (3) of Definition 8.2 decides the equilibrium distance among individuals. Then, the following theorem can be obtained.

Theorem 8.2 (Three-dimensional Cohesive Flocking): *There exists a system consisting of N UAVs governed by* (8.36). *The behavior of each UAV is decided by the protocol* (8.46). *The interaction network of the proposed UAV system is assumed to be undirected and connected at the initial time. Then, the following conclusions hold*

(1) *Each UAV avoids collision with each other.*
(2) *The metric interaction network is always connected.*
(3) *The system asymptotically converge to a consistent configuration that each UAV owns the same speed, the same heading angle, and the same flight-path angle.*
(4) *The system approaches a cohesive configuration that its total potential is minimized.*

In short, we may ignore the time variable in the following proof process for these time-variant variables. For example, we may use N_i instead of $N_i(t)$.

Proof of Theorem 8.2 From $u_i = [n_{ti}, n_{hi}, n_{vi}]^T$ and $H_i = [t_i, n_i, h_i]^T$, control protocol (8.46) can be rewritten as

$$u_i(t) = -\frac{1}{g}H_i\sum_{j\in N_i}(q_i - q_j) + \frac{1}{g}H_iG - \frac{1}{g}H_i^T\sum_{j\in N_i}\nabla_{p_{ij}}V(\|p_{ij}\|) \tag{8.47}$$

As H_i is an orthogonal matrix, $H_i^TH_i = I$. From (8.41) and (8.47), one gets

$$\dot{q}_i = -\sum_{j\in N_i}(q_i - q_j) - \sum_{j\in N_i}\nabla_{p_{ij}}V(\|p_{ij}\|),\ i = 1, \ldots, N \tag{8.48}$$

Let $p = [p_1^T, \ldots, p_N^T]^T$ and $q = [q_1^T, \ldots, q_N^T]^T$. According to (8.41) and (8.48), one gets

$$\dot{q} = -(L_N \otimes I_3)q - \sum_{i=1}^{N}\sum_{j\in N_i}\nabla_{p_{ij}}V(\|p_{ij}\|) \tag{8.49}$$

where I_3 denotes the 3×3 identity matrix, and $L_N(t) = [l_{ij}]_{N\times N}$ is defined as $\xi(t)$'s Laplacian matrix.

Let $\psi = [\psi_1^T, \ldots, \psi_N^T]^T$ and $\gamma = [\gamma_1^T, \ldots, \gamma_N^T]^T$ Consider the following energy function as the common Lyapunov function

$$E(p, q) = \frac{1}{2}V + \frac{1}{2}q^Tq \tag{8.50}$$

Then the derivative of $E(p, q, \psi, \gamma)$ w.r.t time $t \in [t_r, t_{r+1})$ is

$$
\begin{aligned}
\frac{dE}{dt} &= \frac{1}{2}\sum_{i=1}^{N}\sum_{j\in N_i}\dot{p}_{ij}^T\nabla_{p_{ij}}V\left(\|p_{ij}\|\right)+q^T\dot{q}\\
&= \sum_{i=1}^{N}\dot{p}_i^T\sum_{j\in N_i}\nabla_{p_{ij}}V\left(\|p_{ij}\|\right)+q^T(-(L_N\otimes I_3)q-\sum_{i=1}^{N}\sum_{j\in N_i}\nabla_{p_{ij}}V\left(\|p_{ij}\|\right))\\
&= \sum_{i=1}^{N}q_i^T\sum_{j\in N_i}\nabla_{p_{ij}}V\left(\|p_{ij}\|\right)-q^T(L_N\otimes I_3)q-q_i^T\sum_{i=1}^{N}\sum_{j\in N_i}\nabla_{p_{ij}}V\left(\|p_{ij}\|\right)\\
&= -q^T(L_N\otimes I_3)q
\end{aligned}
\tag{8.51}
$$

Undirected graph $\xi(t)$'s Laplacian matrix L_N is symmetric and positive semi-definite, thus we have

$$dE/dt \le 0 \tag{8.52}$$

The potential energy of the system at initial time is finite, and the velocities and attitudes of all UAVs at initial time are also finite. Thus, $E(p(0), q(0))$ (the energy of the system at initial time) is finite. Since $\frac{dE}{dt} \le 0$ on each time-interval $\left[t_r, t_{r+1}\right)$, $r = 0, 1, \ldots$, the supremum of $E(t)$ is obviously its initial value $E(p(0), q(0))$, for all t. Furthermore, both $V(t) = \sum_{i=1}^{N}\sum_{j\in N_i} V\left(\|p_{ij}\|\right)$ and $V\left(\|p_{ij}\|\right)$ are bounded.

Definition 8.2's rule (1) indicates that if $p_{ij} \to 0$, then we have $V\left(\|p_{ij}(t)\|\right) \to \infty$, which violates the above conclusion ($V(p_{ij}(t)$ is bounded). Thus, one gets $p_{ij} > 0$, that is, no collision happens between any two UAVs.

According to rule (2) of definition 8.2, if $p_{ij} \to D$, $(i, j) \in \varepsilon$, then one gets $V\left(\|p_{ij}\|\right) \to \infty$, which is also not consistent with the above conclusion (that is, $V\left(\|p_{ij}\|\right)$ is bounded). Therefore, we have $p_{ij} < D$. A new edge $(i, j) \notin \varepsilon$ is supposed to be added to ε, then the associated potential $V\left(\|p_{ij}\|\right)$ remains finite because $p_{ij}(t) < D-\zeta$. Therefore, the new potential V is finite too. That is to say, the connectivity of the communication network is preserved.

Assume that there are $m_r \in N$ new edges being added to the interaction network at switching time t_r, $r = 1, 2, \ldots$. The UAV system's interaction topology at initial time is supposed to be connected. ξ_1 denotes the interaction topology of the system at the initial time, and ξ_c denotes the set of undirected and connected graphs on the vertices. The sequence of switching topologies ξ_{r+1} within $\left[t_r, t_{r+1}\right)$ consists of such graphs that satisfy $\xi_{r+1} \in \xi_c$. ξ_c is a finite set, since the number of vertices is finite. At most $m \in N$ new edges are assumed can be added to the initial interaction topology ξ_1. It is clear that $0 < m_r \le M$ and $r \le M$, that is, the number of switching time is finite. Consequently, $\xi(t)$ becomes fixed finally. Then, we are going to discuss in further restricted to the time interval $\left[t_f, \infty\right)$, where t_f is the last switching time.

As the distance between neighbors is not longer than D, set $B = \{p, q|E(p, q) \le E(p(0), q(0))\}$ is positively invariant. One gets $p_{ij} < (N-1)D$ for all i and j, due to $G(t)$ is connected for all $t \ge 0$. According to $E(p, q) \le E(p(0), q(0))$, we obtain that $q^T q \le 2E(p(0), q(0))$, that is, $q \le \sqrt{2E(p(0), q(0))}$.

Therefore, the set B is compact (bounded and closed). On the connected time interval $\left[t_f, \infty\right)$, the system (8.36) with control input (8.46) is an autonomous system. Therefore, the trajectories of the UAVs will converge to the invariant set $S = \left\{p, q | \frac{dE}{dt} = 0\right\}$. Besides, $\frac{dE}{dt} = 0$ if and only if $q_i = q_j$ for all $i \in N$ and $j \in N_i$.

According to (8.37), $q_i = q_j$ is equivalent to

$$\begin{cases} \theta_i \cos \gamma_i \cos \psi_i = \theta_j \cos \gamma_j \cos \psi_j \\ \theta_i \cos \gamma_i \sin \psi_i = \theta_j \cos \gamma_j \sin \psi_j \\ \theta_i \sin \gamma_i = \theta_j \sin \gamma_j \end{cases} \tag{8.53}$$

that is, $\theta_i = \theta_j$, $\psi_i = \psi_j$, and $\gamma_i = \gamma_j$.

With $p_{ij} = -p_{ji}$, one gets $\sum_{i=1}^{N} \sum_{j \in N_i} \nabla_{p_{ij}} V\left(\|p_{ij}\|\right) = 0$. From the control input (8.46) and $q_i = q_j$, one has

$$\begin{aligned} \dot{\overline{q}} &= \frac{1}{N} \sum_{i=1}^{N} \dot{q}_i \\ &= \frac{1}{N} \left(-\sum_{i=1}^{N} \sum_{j \in N_i} (q_i - q_j) - \sum_{i=1}^{N} \sum_{j \in N_i} \nabla_{p_{ij}} V\left(\|p_{ij}\|\right) \right) \\ &= \frac{1}{N} \left(-\sum_{i=1}^{N} \sum_{j \in N_i} \nabla_{p_{ij}} V\left(\|p_{ij}\|\right) \right) \\ &= 0 \end{aligned} \tag{8.54}$$

Due to $q_1 = \ldots = q_N = \overline{q}$, we have $\dot{q}_i = \dot{\overline{q}} = 0$, that is, $gH_i^T u_i + G = 0$, $i = 1, \ldots, N$. Thus, we obtain that $-\sum\limits_{j \in N_i} (q_i - q_j) - \sum\limits_{j \in N_i} \nabla_{p_{ij}} V\left(\|p_{ij}\|\right) = 0$ for $i = 1, \ldots, N$. With $q_i = q_j$, one gets $\sum\limits_{j \in N_i} \nabla_{p_{ij}} V\left(\|p_{ij}\|\right) = 0$. The minimum value of V is obtained by

$$\begin{aligned} \frac{dV}{dt} &= \sum_{i=1}^{N} \dot{p}_{ij}^T \sum_{j \in N_i} \nabla_{p_{ij}} V\left(\|p_{ij}\|\right) \\ &= 2 \sum_{i=1}^{N} \dot{p}_i^T \sum_{j \in N_i} \nabla_{p_{ij}} V\left(\|p_{ij}\|\right) \\ &= 0 \end{aligned} \tag{8.55}$$

So far, the four conclusions of Theorem 8.2 are all proved.

8.2.3 Results

Take 80 UAVs as an example, numerical simulations are carried out in MATLAB to verify the four conclusions of Theorem 8.2. Furthermore, flocking scenarios of these 80 UAVs under different interaction requirements are quantitatively analyzed and compared statistically from the point of energy and time savings. After that, the collective effects under different group sizes, that is, the scalability effects, are discussed in further.

8.2.3.1 Initial Conditions

N UAVs are supposed to fly under free boundary conditions. Each UAV is steered by (8.36). The initial position of each UAV is randomly generated in a cubic area with size L, and its initial speed is randomly generated from the interval $(0, 0.5)$. Besides, its initial heading angle and flight-path angle are randomly generated from the interval $\left(-\frac{\pi}{2}, \frac{\pi}{2}\right)$.

To keep the continuity and comparability of simulation results under different group sizes, these UAV systems' initial density $\rho = \frac{N}{L^3}$ should be a constant. That is to say, the scale of group size N directly proportional to the cube of the area size L. In the following simulations, the initial density $\rho = 0.1$, and each experiment's simulation time $T = 100\,\text{s}$.

The specific potential function in the control protocol (8.46) is given as [49]

$$V\left(\|p_{ij}(t)\|\right) = \frac{b}{\|p_{ij}(t)^2\|} - aln\left(D^2 - \|p_{ij}(t)\|^2\right) \tag{8.56}$$

where the first part is the repulsive potential item and the second part is the attractive potential item. Here, a is an attractive coefficient, and b is a repulsive coefficient. There exists a unique distance $\|d^*\| = \sqrt{\frac{\sqrt{b^2+4abD^2}-b}{2a}}$, at which the attraction and repulsion balance. Without loss of generality, the parameters are chosen as $a = 1$, $b = 1$, and thus $\|d^*\| = \sqrt{\frac{\sqrt{1+4D^2}-1}{2a}}$.

One of the most important initial conditions is that the initial communication network must be connected. To satisfy this condition, three steps should be carried out. First, the initial positions of these UAVs are generated randomly. Second, the connectivity of the communication network based on the above initial positions is checked. If the network is connected, then these UAVs implement the simulations under control protocol (8.46); otherwise, a new set of initial positions is randomly generated for these UAVs. Third, the above process is repeated until a connected initial communication network is found out.

8.2.3.2 Simulation Example

In order to validate the effectiveness of these conclusions of Theorem 8.2, $N = 80$ UAVs are taken as an example to execute the cohesive flocking task. Each UAV governed by (8.36) adjusts its behavior under the control protocol (8.46). The communication radius D is 4 m.

Figure 8.10 gives the readers a detailed intuition into the results of the proposed simulation. As shown in Fig. 8.10a, 80 UAVs start flying from these green star points, and their final positions due to the simulation time T = 100 s is expressed by these colorful balls. Each curve denotes the trajectory of one UAV The curves with different colors denote the trajectories of different UAVs. Figure 8.10a has given a clear exhibition that 80 UAVs asymptotically converge to flying in a flock. Eighty UAVs are considered in this case, thus there are $N(N-1)^2 = 80 \times 792 = 3160$ curves in Fig. 8.10b. Each curve denotes the distance between two UAVs. According to the simulation result shown in Fig. 8.10b, the distances between any two UAVs finally become stable. During the evolutionary process, the distances between any two UAVs are all larger than 0, that is to say, there is no collision between any two UAVs. More importantly, the interference among UAVs can be reduced in terms of adjusting the repulsive term, such as increasing the value of repulsive coefficient b. Figure 8.10c–e, respectively, gives the velocities, the heading angles, and the flight-path angles of 80 UAVs versus time.

The above simulation results have definitely proved that these UAVs will form a cohesive and consistent configuration without collisions between any two UAVs.

8.2.4 Discussion

According to the above analysis, we already know that the UAV system will converge to a flock finally, on condition that the initial interaction network is connected. Then, a more important question is coming, that is, how to accelerate the emergency of a flock. In the following paragraphs, the energy and time costs, as well as scalability effects of emerging a flock are evaluated from the point of statistical significance and strength quantitatively.

To guarantee that all of the UAVs will converge to cohesive flocking finally, one condition must be satisfied, that is, the initial interaction topology of the UAVs should be connected. As the nearest-neighbor interaction rules have been adopted and every UAV has the same communication radius, the common communication radius becomes the only factor that can change the network connectivity. The value of the communication radius decides the scale of the interaction network and also the neighbor number of each UAV.

The initial state of the UAV system is supposed to be given. Let $D_{\min}$ denote the minimal value of the communication radius that forces the interaction network to be connected. For these connected communication radii $D \in [D_{\min}, \infty)$, a smaller value means fewer neighbors for each UAV. This situation further causes the decrease of

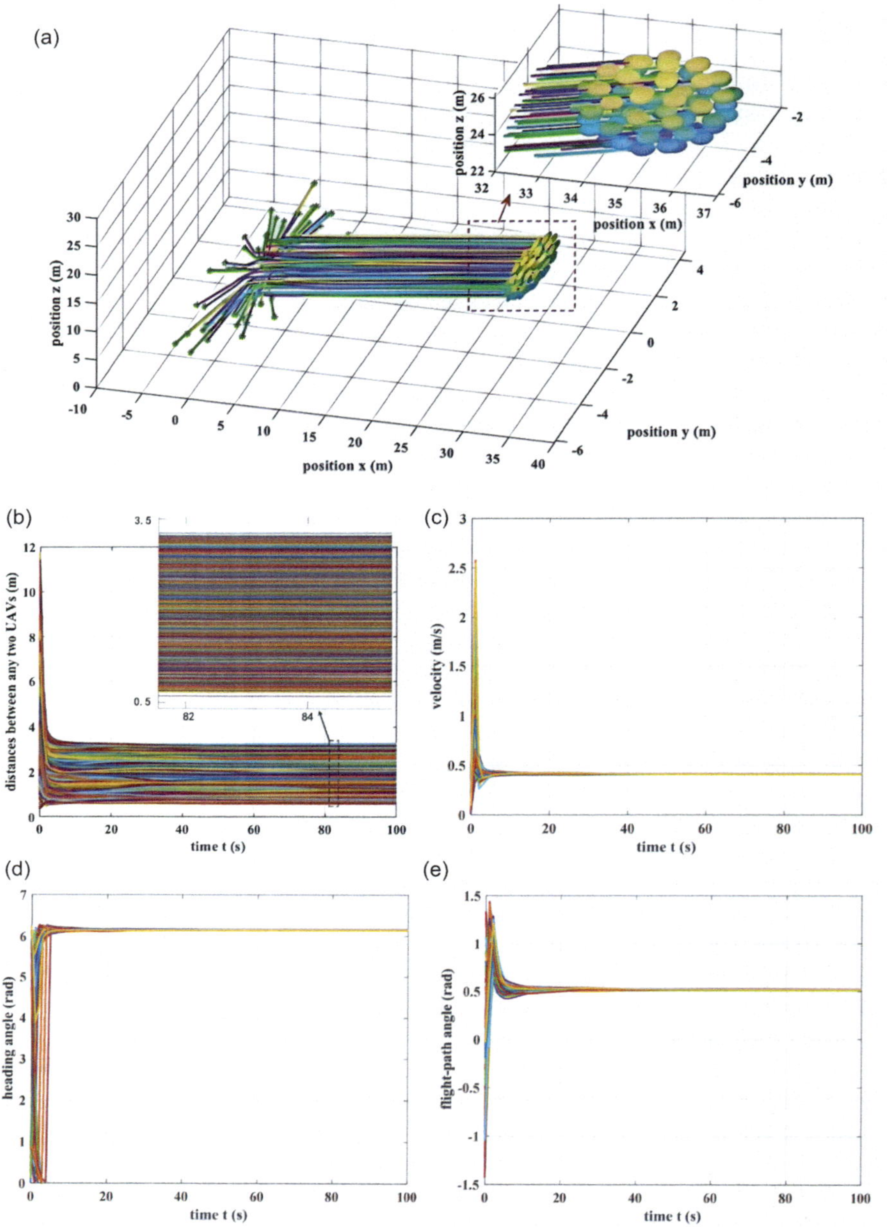

Fig. 8.10 Simulation results for three-dimensional collective cooperation of \$80\$ UAVs. **a** trajectories of \$80\$ UAVs, and details of the final configuration; **b** distances between any two UAVs versus time; **c** speeds of 80 UAVs versus time; **d** heading angle of 80 UAVs versus time; **e** flight-path angle of 80 UAVs versus time

computational complexity and calculated quantity of the whole system, as well as the reduction of energy consumption.

The distance between any two UAVs is one of the proper variables to draw the convergence speed of UAV flocking and also reflects the stability of the UAV system. If the distance between any two UAVs becomes stable, then the system becomes stable. Therefore, energy measurements can be evaluated simply by the following order parameter

$$EE(t) = \frac{\sum_{i=1}^{N-1} \sum_{j=i+1}^{N} d_{ij}(t)}{N(N-1)/2} \tag{8.57}$$

where $d_{ij}(t) = \|p_i(t) - p_j(t)\|, i, j \in v$.

To assess the energy and time costs of a flock, numerical simulations of \$80\$ UAVs are carried out under a series of connected communication radii. There are mainly three steps: First, the initial states of 80 UAVs are produced randomly. Second, connected communication radius' sample values are linearly selected from the interval [2.8, 8]. Finally, starting from the same initial states of 80 UAVs, the experiment is carried out under each sample communication radius. Figure 8.11b shows the specific potential functions under these sample communication radii, and the simulation results are shown in Fig. 8.11a. Different colors describe the simulation results under different sample communication radii. It is clear to see that the system converges to a flock faster under a smaller communication radius, although in which a longer journey (equaling to the initial individual distance minus the stable individual distance) should be covered to reach the stable individual distance $d^* = \sqrt{D}$.

Moreover, the above conclusion is obtained according to a fixed initial state and fixed group size. In order to make the conclusion more rigorous and more precise, each experiment is repeated over 100 trials that start with different initial states. Convergence time t_c is adopted as the order parameter to evaluate the time cost of

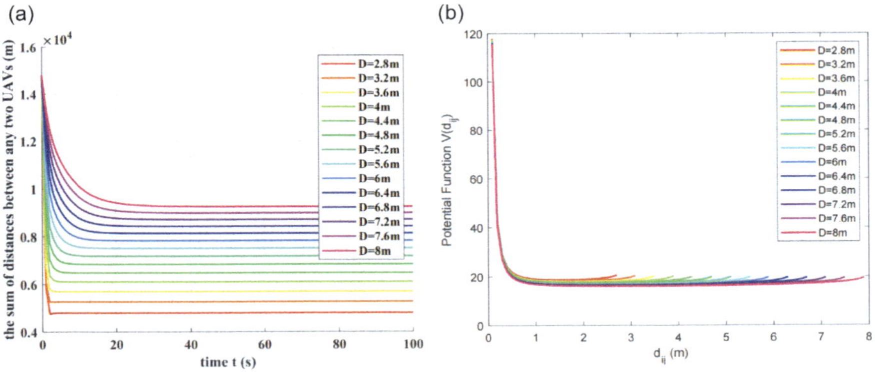

Fig. 8.11 Energy costs of 80 UAVs versus time under different connected communication radii and relative specific functions. **a** energy costs; **b** specific functions

each experiment. The definition of convergence time t_c is the minimal value of t_k, which satisfies the following constraint condition

$$\frac{EE(t_k) - EE(T)}{EE(T)} \leq K \tag{8.58}$$

where $K = 0.01$. The standard error bar is defined as

$$\sigma = \frac{1}{\sqrt{n}} t_c \tag{8.59}$$

where $n = 100$ is the experiment number.

The relationships between convergence time and sample communication radius under different group sizes are shown in Fig. 8.12. In all of these figures, the system converges to a flock within simulation time. For each given group size, the convergence time increases with the enlargement of communication radius most of the time when $D \geq D_{\min}$. That is to say, under the premise that the initial communication network is undirected and connected, the minimal communication radius causes the largest convergent speed. The minimal communication radius means the minimal edge set of the connected interaction network of the UAV system, at the same time, requires minimal computation capability for each UAV during the whole evolutionary process. This conclusion can be easily used to extend the distributed control algorithm from small groups to large-scale groups of UAVs and to skillfully avoid the awkwardness of exponential growth caused by the increase in group size.

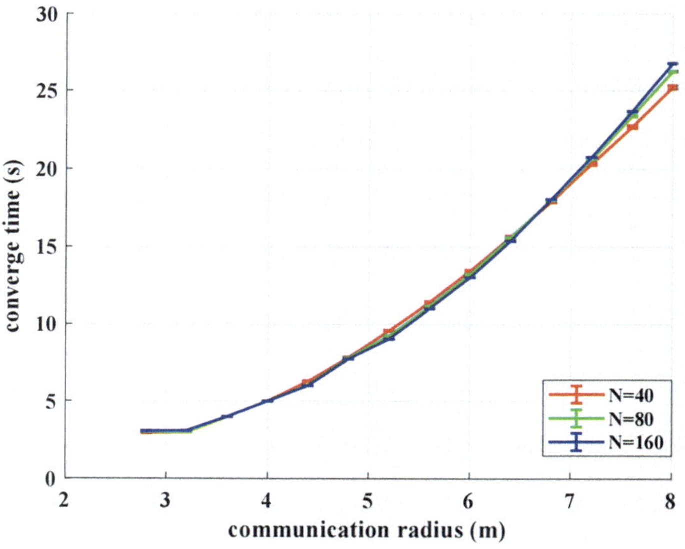

Fig. 8.12 Time cost of flocking system (8.36)

Scalability performance is examined by increasing the robot group size. In this case, communication radius D is fixed to 4 m. The energy cost $EE(t)$ is drawn versus time when the number of UAVs is double increased from 40 to 160. Figure 8.13 shows the good scalability performance of the flocking system (8.36) under control protocol (8.46), as the converge time almost the same following with the increase of group size. I have labeled the coordination value of three points on the three curves. There are $N(N-1)/2$ edges between any two UAVs for a system consisting of N UAVs. When $N = 40$, then we have $N(N-1)/2 = 780$ edges, and the average distance between any two UAVs is $1493.6695/780 = 1.915 \approx 2$. Similarly, when $N = 80$, then we have $N(N-1)/2 = 3160$ edges, and the average distance between any two UAVs is $6107.8093/3160 = 1.93 \approx 2$; when $N = 160$, then we have $N(N-1)/2 = 12,720$ edges, and the average distance between any two UAVs is $24,788.5314/12,720 = 1.95 \approx 2$. They all equal to the sqrt of communication radius D. Thus, Fig. 8.13 have proved the conclusion that the stable individual distance $d^* = \sqrt{D}$. Furthermore, Fig. 8.13 reinforces this conclusion that the converge times for different group sizes are almost the same when the value of the connected communication radius is small enough.

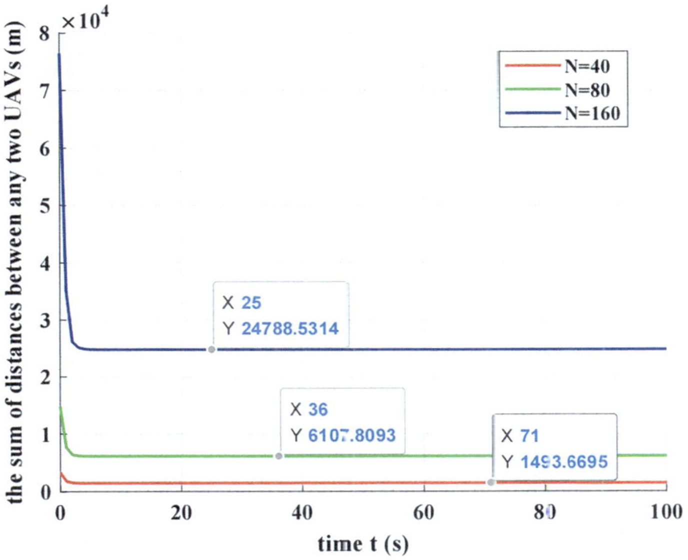

Fig. 8.13 Scalability effect of flocking system (8.36) under control protocol (8.46)

8.2.5 Summary

In conclusion, we have introduced a systematically and logically rigorous research methodology for studying these interdisciplinary topics, such as aerial swarms. In this paper, we mainly focus on combing the control theories and statistical analysis of simulation results to solve the optimization of the emergence of order in a flock. The UAV system is supposed to be an egalitarian system that obeys the nearest-neighbor interaction rules. Therein, each UAV is depicted by a simplified mathematical model with nonlinear constraints. Theoretical analysis of the UAV flocking steered by a distributed control algorithm is provided, as long as the initial communication network is connected. Communication radius is considered as the only scale to draw the interaction topology, a statistical method is therefore proposed to scrutinize the relationship between communication radius of each UAV and convergence speed of these UAVs based on the above theoretical framework. Convergence time serves as a criterion to evaluate the convergence effect in the flock. Statistical analysis of abundant simulation results is provided to optimize the emergence of collective behavior by evaluating energy/time measurements and scalability effects. Results show that smaller communication radius causes faster convergence of the aerial group. To some extent, this conclusion echoes from the biological evidence that each starling interacts with six or seven neighbors most often at the same time. More importantly, this conclusion offers empirical evidence on the transition from scale-dependent interaction to topology-dependent interaction.

8.3 The Comparison Between Egalitarian and Hierarchical Systems

We present a general framework for modeling a wide selection of flocking scenarios under free boundary conditions. Several variants have been considered - including examples for the widely observed behavior of hierarchically interacting units. The models we have simulated correspond to classes of various realistic situations. Our primary goal was to investigate the stability of a flock in the presence of noise. Some of our findings are counter-intuitive in the first approximation, e.g., if the hierarchy is based purely on dominance (an uneven contribution of the neighbors to the decision about the direction of flight of a given individual) the flock is more prone to loose coherence due to perturbations even when a comparison with the standard egalitarian flock is made. Thus, we concentrated on building models based on leader–follower relationships. And, indeed, our findings support the concept that hierarchical organization can be very efficient in important practical cases, especially if the leader–follower interactions (corresponding to an underlying directed network of interactions) have several levels. Efficiency here is associated with remaining stable (coherent and cohesive) even in cases when collective motion is destroyed by random perturbations. The framework we present allows the study of several further complex interactions among the members of flocking agents.

8.3.1 Introduction

Collective behavior is a very important aspect through which small or large groups of organisms optimize their living [68]. It involves collective decision making various contexts, such as searching for food, navigating towards a distant target [69–72] or deciding when and where to go [73–75]. Flocking is perhaps the most common and spectacular manifestation of collective behavior not only in nature since recently has gained attention in the context of collective robotics as well [76–78]. Most of the experimental and modeling approaches aimed at describing flocking by assuming egalitarian interactions among the members of a flock.

However, just as flocking is a widespread behavioral pattern of a collective, the hierarchical structure of the interactions among the members of groups is also very common [79]. Thus, starting with a trend-setting paper of Couzin et al. [80] the question of leadership during flocking has attracted increasing interest. Early works assumed a two-level hierarchy while recent experimental observations involving some sophisticated animal groups such as pigeons or primates point towards the possibility of significantly more complex internal organization principles during group motion [69, 75, 81, 82]. In socially highly organized groups beyond a given size (dozens or so) the roles related to leadership do not seem to be simply binary, but several levels of hierarchy can be identified. In particular the available few experimental results indicate that many co-moving groups have an internal system of interactions which can be best interpreted in terms of pairwise hierarchical levels of interactions. Perhaps the best quantitatively evaluated hierarchical group motion was carried out for pigeon flocks [69, 83] in which a hierarchically distributed set of interactions was demonstrated using GPS tracks and a velocity correlation-delay method [69].

On the other hand, in spite of its relevance, there have been no realistic models proposed and studied devoted to the understanding of the conditions under which hierarchical flocking can be optimal. The only vaguely related original model is in a beautiful work by Shen [84], but the assumptions are quite arbitrary and the effect of the hierarchy on the performance of the flock is not investigated.

Therefore, our goal here is to design a general framework which allows the treatment of questions related to hierarchical leader–follower interactions in a general setting, when the flock is freely moving (in the present study in two dimensions). We also discuss the various variants allowed by our approach and characterize the efficiency of the behavioral rules by determining to what level a flock is stable against external perturbations. Hierarchy is thought to be prevalent because it can be shown to result in more efficient group performance [75, 79].

The present work is the first one in which a flexible and plausible framework is introduced to model hierarchical flocking. It turns out that introducing pairwise interactions satisfying a realistic hierarchical dynamics is far from being trivial. After introducing the model the main aspect we investigate is the stability of a flock moving freely in a border-less two-dimensional space. We shall associate with stability the tendency of a flock not to break into parts under external perturbations.

8.3.2 Definitions Used in Our Model

For an egalitarian system, every particle is believed to be the same, while for a hierarchical system, individual difference exists. Any complex multi-agent system can be classified as egalitarian or hierarchical according to the set of pairwise relationships between any two individuals. For each pair of particles in an egalitarian system, both of them have the same ability (referring to the same contribution strength and bidirectional information flow) when making decisions for the next time step. Meanwhile, for each pair of particles in a hierarchical system, when making a decision, they may have different level of contribution to the decision (weight) or have a directed information flow relationship, like the leader–follower mechanism (follower has no influence on the behavior of the leader).

Graphs/networks represent a useful tool for visualizing these pairwise interaction relationships of individuals belonging to one co-moving group. Therefore, we define several matrices to describe the internal properties of the hierarchical mechanisms from different points of view, for the purpose of characterizing the differences between the egalitarian and hierarchical systems clearly.

8.3.2.1 Contribution Matrix

Contribution matrix $C_N = [c_{ij}]_{N\times N}\left(c_{ij} > 0\right)$ is defined to describe the contribution strength (weight) of each particle during the decision making process regarding the new preferred directions of the particles.

(1) For an egalitarian system, every particle has the same contribution value, that is, $c_{ij} = q(q > 0)$, for $i, j = 1, ..., N$. q is a constant.
(2) For a hierarchical system, not all of these $c_{ij}(i, j = 1, ..., N)$ have the same positive value. For example, $c_{ij} = q(q > 0)$, for $i, j = 1, ..., N$, or c_{ij} satisfies some probability distribution(such as log-normal), for $i, j = 1, ..., N$. We name this kind of hierarchy as contribution driven hierarchical system.

8.3.2.2 Dominance Matrix

The dominance matrix $B_N = [b_{ij}]_{N\times N}$ is defined to describe the direction of information flow between each pair individuals. The direction of information flow specifies the set of particles whose behavior influences the decision of a given particle at each time step.

(1) For each pair of particles i and j $(i \neq j)$ in an egalitarian system, their behaviors can be influenced by each other, that is to say, the information between each pair of particles is transmitted bidirectionally (corresponding to an undirected graph). Thus, in an egalitarian system for each pairwise interaction $b_{ij} = b_{ji} = 1, \forall i, j = 1, ..., N, i \neq j$.
(2) If the information flow of paired particles is directional, that is, only one particle can obtain the other particle's information (directed graph), then we have a dominance driven mechanism, which is another kind of hierarchy named dominance hierarchical system. For paired particle i and j, if i is led by j, then we have $b_{ij} = 1$ and $b_{ji} = 0$ $(\forall i, j = 1, ..., N, i \neq j)$.
(3) If $b_{ii} = 1(i = 1, ..., N)$, then B_N is called dominance matrix. If $b_{ii} = 0(i = 1, ..., N)$, then B_N is called strictly dominance matrix. We call the egalitarian system with $b_{ii} = 0(i = 1, ..., N)$ as strictly-egalitarian system. The egalitarian system discussed in this paper refers to the case that $b_{ii} = 1(i = 1, ..., N)$. We use the terminology dominance hierarchical system with $b_{ii} = 0(i = 1, ..., N)$ as dominance strictly-hierarchical system.

Leader–follower mechanism is a typical kind of dominance relation in hierarchical organizations. For each pair of particles, leader particle does not take into account the influence of the follower, but the follower considers the behavior of the leader particle when making decision on its behavior at the next step. This feature is represented by the fact that the matrix B_N is not symmetric. Without loss of generality, we number these particles according to the level of dominance from 1 to N. Particle 1 is the strongest one of the whole system, while particle N is the weakest one. Therefore, the dominance matrix B_N is a complete and symmetric matrix for egalitarian systems; for a dominance hierarchical system, with non-zero diagonal elements is a lower-triangular matrix; for dominance strictly-hierarchical system, B_N is a strictly lower-triangular matrix (strictly triangular matrix refers to the triangular matrix whose diagonal elements are all zero).

8.3.2.3 Egalitarian Versus Hierarchical Systems

Now we can give a formal definition of egalitarian and several kinds of hierarchical systems.

(1) **Egalitarian system**

For each pair of individuals, if both of them will use each other's information with the same weight to decide the behavior at the next step, we say it is an egalitarian system. An egalitarian system satisfies the following two rules:

(i) $c_{ij} = q, \forall i, j = 1, ..., N$, q is a positive constant;
(ii) $b_{ij} = 1, \forall i, j = 1, ..., N$.

(2) **Contribution driven hierarchical system**

A contribution hierarchical system satisfies the following two rules:

(i) c_{ij} follows some distribution for $i, j = 1, \ldots, N$ (thus, not all of these c_{ij} have the same positive value);
(ii) $b_{ij} = 1, \forall i, j = 1, \ldots, N$.

(3) **Single-layer leader–follower hierarchical system (dominance driven hierarchical mechanism)**

A single-layer leader–follower hierarchical system satisfies the following two rules:

(i) $c_{ij} = q, \forall i, j = 1, \ldots, N$, q is a positive constant;
(ii) $b_{ij} = 1, i \geq j, i, j = 1, \ldots, N$.

(4) **Single-layer leader–follower strictly-hierarchical system (dominance driven strictly-hierarchical mechanism)**

A single-layer leader–follower strictly-hierarchical system satisfies the following two rules:

(i) $c_{ij} = q, \forall i, j = 1, \ldots, N$,q is a positive constant;
(ii) $b_{ij} = 1, i > j, i, j = 1, \ldots, N$.

(5) **Double-layer leader–follower hierarchical system (contribution driven dominance hierarchical mechanism)**

In case the weights in the contribution matrix are not equal, we associate the system with the presence of dominance (the contribution of agents having a larger weight dominate over the contribution by those with a smaller weight). We consider such systems whose behavior is determined through a *contribution driven* mechanism. Then, the particle with larger weight contribution is named as leader, while the other one is named as follower. A double-layer leader–follower hierarchical system satisfies the following two rules:

(i) c_{ij} follows some distribution for $i, j = 1, \ldots, N$ (but not all of these c_{ij} have the same positive value);
(ii) $b_{ij} = 1, i \geq j, i, j = 1, \ldots, N$.

(6) **Double-layer leader–follower strictly-hierarchical system (contribution driven dominance strictly-hierarchical mechanism)**

A double-layer leader–follower strictly-hierarchical system satisfies the following two rules:

(i) c_{ij} meets some distribution for $i, j = 1, \ldots, N$ (but not all of these c_{ij} have the same positive value);
(ii) $b_{ij} = 1, i \geq j, i, j = 1, \ldots, N$.

The above systems can be characterized by their intersection matrix $\$c_{ij} * b_{ij}$, $i, j = 1, ..., N$. For an egalitarian system, it is a complete matrix with the same elements. For contribution hierarchical systems, it becomes a complete matrix with various elements. For single-layer leader–follower hierarchical system, it is a lower-triangular matrix with the same lower-triangular elements. For single-layer leader–follower strictly-hierarchical system, it is a strictly lower-triangular matrix with the same lower-triangular elements. For double-layer leader–follower hierarchical system, it becomes a lower-triangular matrix with varying lower-triangular elements. For double-layer leader–follower strictly-hierarchical system, it becomes a strictly lower-triangular matrix with varying lower-triangular elements.

Besides, in a more general model, c_{ij} and b_{ij} can be time-dependent. If c_{ij} and b_{ij} is not time-dependent that means everyone in the system has fixed a set of relationships. If $c_{ij}(t)$ and $b_{ij}(t)$ are time-dependent, then the relationships among these particles change with time. Other different variants can thus be defined according to the contribution matrix $c_{ij}(t)$ and dominance matrix $b_{ij}(t)$. Here, we only discuss the case that when c_{ij} and b_{ij} are constant.

8.3.3 *Hierarchical Model for Flocking*

We consider in this paper N particles moving continuously (off lattice) in a free area without any boundary limitation. As shown in Fig. 8.14, the position and direction of N particles at the beginning are generated randomly, while over time these particles are expected to move coherently (ordered state). The figure is for the noiseless case.

Suppose that the time interval between two updates of the directions and positions is $\Delta t = 1$. This assumption can be made without loosing generality since Δt occurs only in combination (being multiplied by it) with velocity terms. At $t = 0$, N particles were randomly distributed within an area of a given size and have the same absolute

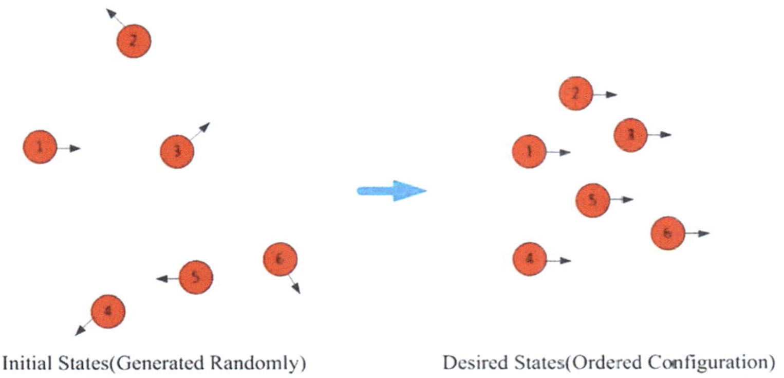

Fig. 8.14 Initial configuration versus the ordered phase of collective motion

velocity v_d as well as randomly distributed directions θ. At each time step, the position of the ith particle is updated according to

$$x_i(t+1) = x_i(t) + v_i(t) \tag{8.60}$$

In each time step, the velocity of a particle $v_i(t+1)$ is updated according to the following equation

$$v_i(t+1) = v_i^{align}(t) + v_i^{rep}(t) + v_i^{adh}(t) \tag{8.61}$$

where $v_i^{align}(t)$ is the alignment term, $v_i^{rep}(t)$ is the repulsion term, and $v_i^{adh}(t)$ is the attraction term.

The alignment term was constructed to have an absolute value ϑ and a direction given by the angle $\theta_i^{align}(t+1)$. The angle was obtained from the expression

$$\theta_i^{align}(t+1) = \langle\theta_i(t)\rangle + \Delta\theta(t) \tag{8.62}$$

where $\Delta\theta(t)$ represents noise, which is a random number chosen with a uniform probability from the interval $[-\eta/2, \eta/2]$. $\langle\theta_i(t)\rangle$ denotes the average direction of the velocities of neighbors of the given particle \$i\$. The average direction is given by the angle

$$\langle\theta_i(t)\rangle = \arctan\left(\frac{\sum_{j=1}^{N} l_{ij}(t)\sin(\theta_j(t))}{\sum_{j=1}^{N} l_{ij}(t)\cos(\theta_j(t))}\right) \tag{8.63}$$

Neighbor matrix $L_{ij}(t) = [l_{ij}(t)]_{N\times N}$ describes the neighbor relationships of particles at time t, where

$$l_{ij}(t) = c_{ij} * b_{ij} * a_{ij}(t), \forall i, j = 1, ..., N \tag{8.64}$$

The definition of adjacency matrix $A_N(t) = [a_{ij}(t)]_{N\times N}$ is

$$a_{ij}(t) = \begin{cases} 1, i = 1, ..., N, j \in N_i(t) \\ 0, otherwise \end{cases} \tag{8.65}$$

where $N_i(t) = \{j \mid \|x_i(t) - x_j(t)\| \leq r\}$. Here, r denotes the interaction radius. Using the above expressions the alignment term can be written as

$$v_i^{align}(t+1) = c^{align} v e_i(t) \tag{8.66}$$

where c^{align} is the coefficient of the alignment term. $e_i(t)$ is a unit vector with direction angle $\theta_i^{align}(t)$.

The repulsion term exists only when the distance between any two particles is smaller than the repulsive radius r_{rep}. And the repulsion term is defined as

$$v_i^{rep}(t+1) = c^{rep} \sum_{j=1}^{N} l_{ij}(t) \left(\frac{r_{rep} - \|x_{ij}(t)\|}{r_{rep}} \cdot \frac{x_{ij}(t)}{\|x_{ij}(t)\|} \right) \tag{8.67}$$

where $\|x_{ij}(t)\| < r_{rep}$ and c^{rep} is the coefficient of the repulsion term.

The attraction term is only considered for the boundary particles [85] of the whole system when the distance between two particles is between r_{rep} and r_{adh}.

$$v_i^{adh}(t+1) = c^{att} \sum_{j=1}^{N} l_{ij}(t) \left(\frac{r_{rep} - \|x_{ij}(t)\|}{r_{att} - r_{rep}} \cdot \frac{x_{ij}(t)}{\|x_{ij}(t)\|} \right) \tag{8.68}$$

where $r_{rep} \leq \|x_{ij}(t)\| \leq r_{adh}$ and c_{adh} is the coefficient of the attraction term. This term is introduced in order to prevent the flock spreading (or, in other words, "evaporating") due to perturbations.

Figure 8.15 shows the neighbour matrix of several variants of flocks mentioned in the last section. From Fig. 8.15, visualizes the differences between the egalitarian system from the typical hierarchical systems.

8.3.4 Simulations and Discussion

The simulations were carried out in a free two-dimensional area. We considered groups of particles having various sizes ranging from 10 to 1280 particles. In order to keep the continuity and comparability of these simulation results of different group sizes, we chose the scale of the area for the random initial positions directly proportional to the scale of group size N, and generated the initial angle from the interval $(-\pi, \pi)$. Here, without loss of generality, the random initial positions are generated from a rectangle area with length L and width $0.6L$. Then, the initial density of the system $\rho = N/(0.6L^2)$, which is constant for various group sizes. We aimed at comparing the stability of egalitarian systems versus various hierarchical systems for varying levels of external disturbances (noise) and for different group sizes.

The alignment item of egalitarian flock model is much like the self propelled particle model proposed by Vicsek et al. in 1995 [86], where the definition of neighbors of particle i includes itself and the contribution of the particles is the same. Thus, we call the egalitarian model as VEM (E for egalitarian, M for model). The contribution driven hierarchical model is called as CHM (C for contribution).

According to the definition of the strictly dominance matrix B_N, the neighbor matrix of a dominance strictly-hierarchical system has zero-value diagonal elements, that is, we have $l_{ii} = 0, \forall i = 1, ..., N$ for all dominance strictly-hierarchical systems. $l_{ii} = 0$ means that the neighbor set of particle i doesn't contain itself. In the

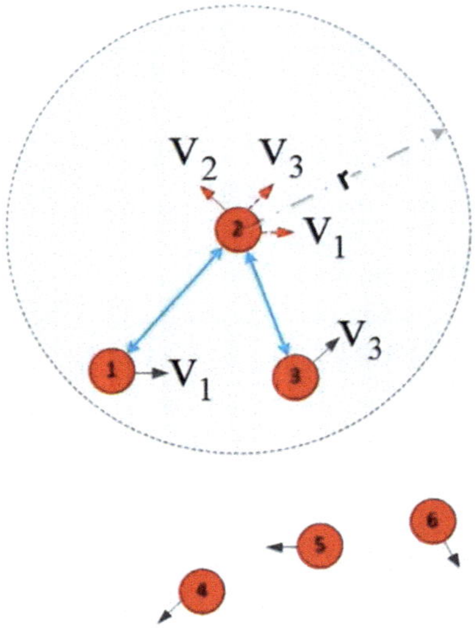

(a) $c_{ij} = q$, $b_{ij} = 1$, $\forall i, j$

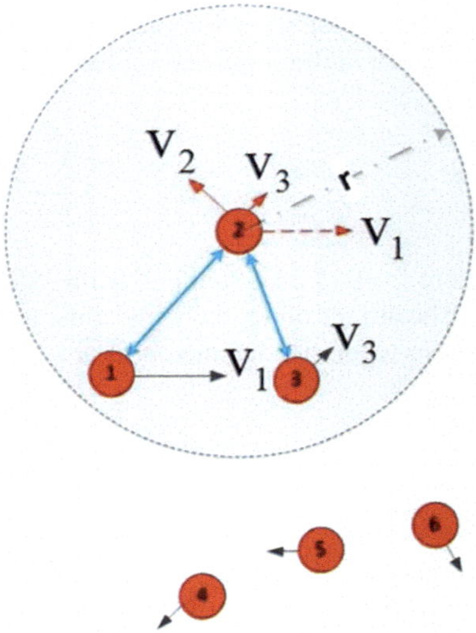

(b) c_{ij} satisfies some distribution, $b_{ij} = 1$, $\forall i, j$

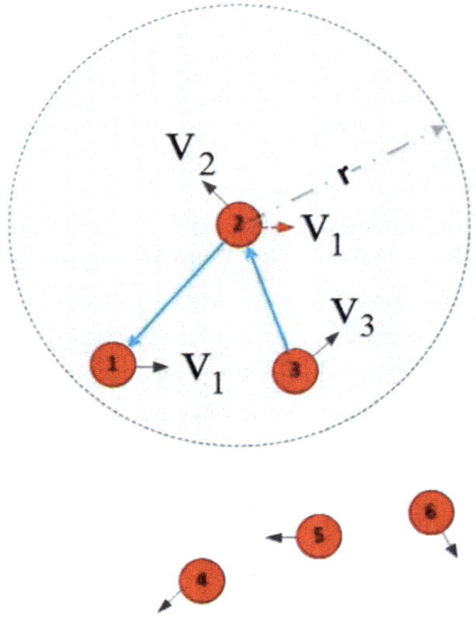

(c) $c_{ij} = q$, $b_{ij} = 1$, $\forall i, j$, $i > j$

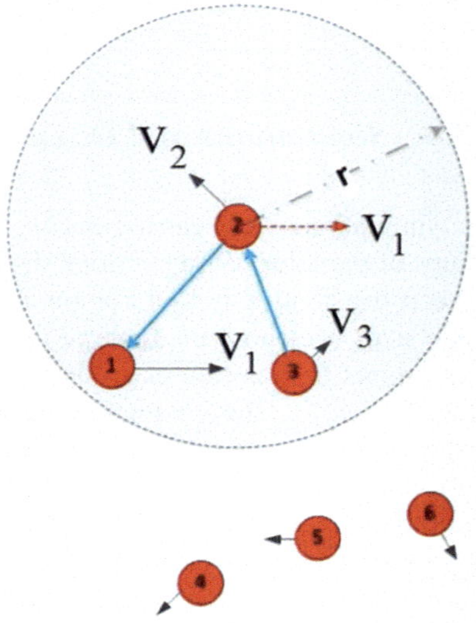

(d) c_{ij} satisfies some distribution, $b_{ij} = 1$, $\forall i, j$, $i > j$

Fig. 8.15 Neighbor matrix. **a** egalitarian system; **b** contribution driven hierarchical system; **c** single-layer leader–follower strictly-hierarchical system; **d** double-layer strictly-hierarchical system. For each figure, only those velocity vectors which are marked by red arrows will be used to calculate the velocity of particle 2 at the next step

following under hierarchy, we always mean hierarchical leader–follower kind of hierarchy. Therefore, we name the single-layer leader–follower hierarchical model ($l_{ii} \neq 0, \forall i = 1, ..., N$) as SHM (S for single-layer leader–follower), and we call the single-layer leader–follower strictly-hierarchical model by SHMZ (S for single-layer leader–follower, and Z for $l_{ii} = 0, \forall i = 1, ..., N$). Similarly, the double-layer leader–follower strictly-hierarchical flock model ($l_{ii} = 0, \forall i = 1, ..., N$) is named as 'Double Hierarchical Model with Zero-s (DHMZ)', while when $l_{ii} \neq 0, \forall i = 1, ..., N$, we call it 'Double Hierarchical Model (DHM)'.

8.3.4.1 VEM and CHM/SHM

In these simulations, we used $\rho = 0.5$, $v = 0.1$, $r_{rep} = 0.1$, $r_{adh} = 1.1$, $c^{align} = 1$, $c_{rep} = 0.1$, $c^{adh} = 0.0002$, and interaction radius $r = 1.1$.

(1) **Order Parameter**

For this case we used the average normalized velocity as the order parameter, defined as

$$\phi^{ave} = \frac{1}{T}\frac{1}{N}\int_0^T \left\| \sum_{i=1}^{N} v_i(t) \right\| dt \tag{8.69}$$

where $T = 2000$ is the simulation time for each experiment.

The error bar is defined as:

$$\sigma = \frac{1}{\sqrt{n}}\overline{\phi} \tag{8.70}$$

where $\overline{\phi}$ is the standard deviation of the order parameter values, and $n > 0$ is the number of simulations for a given system size. The typical values of n were chosen as follows:

$$n = \begin{cases} 1000, when\, N = 10 \\ 1000 when\, N = 20 \\ 500, when\, N = 40 \\ 200, when\, N = 80 \\ 100, when\, N = 160 \end{cases} \tag{8.71}$$

(2) **Simulation Results Analysis and Discussion**

First, we compared the stability of VEM and some simple hierarchical flocking systems, such as CHM and SHM. According to our results (see Figs. 8.16 and 8.17), a simple dominance based system does not perform better than the much studied egalitarian flock. It should be noted that we did not compare VEM with CHM/SHM

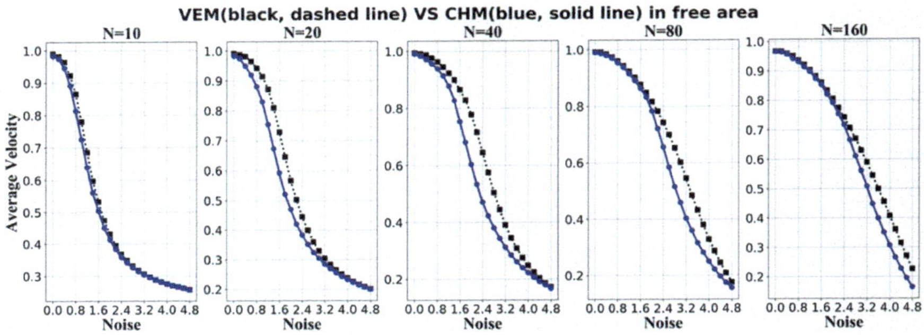

Fig. 8.16 Quantitative comparison of VEM and CHM ($\boldsymbol{l_{ii}} \neq 0, \forall \boldsymbol{i} = 1, \ldots, \boldsymbol{N}$) for various group sizes. c_{ij} satisfies log-normal distribution (whose mean is 0 and standard deviation is 1) for the CHM system. Note that both Fig. 8.16 and Fig. 8.17 show that the egalitarian system is more stable than the simple dominance-based (referring to contribution) hierarchical system

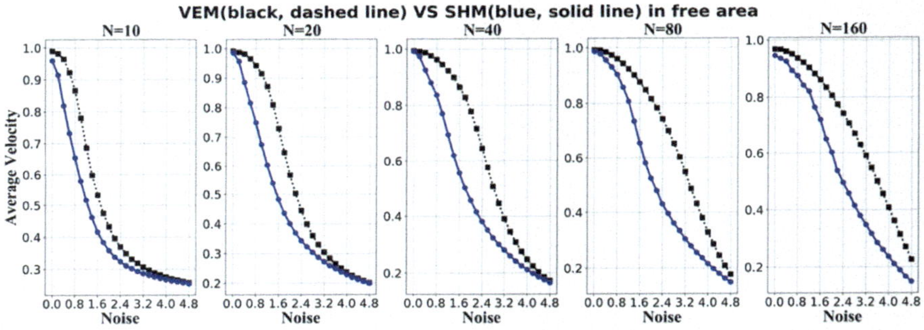

Fig. 8.17 Quantitative comparison of VEM and SHM ($c_{ij} = q(q > 0)$) for various group sizes

for larger group sizes than $N = 160$, because we guess CHM/SHM still do not perform better than VEM, but readers who are interested in size dependence are welcome to investigate it.

8.3.4.2 VEM and DHM/DHMZ

In this simulation, we use $\rho = 0.4$, $\upsilon = 0.1$, $r_{rep} = 0.5$, $r_{adh} = 2.2$, $c^{align} = 1$, $c^{rep} = 0.1$, $c^{adh} = 0.00006$, and interaction radius $r = 2.2$.

(1) **Order Parameter**

As mentioned above, we shall associate the stability with the tendency of a flock not to break into parts under external perturbations. In order to measure the stability of the particle group, we used the following velocity correlation as the order parameter

$$\phi^{corr} = \frac{1}{T}\frac{1}{N(N-1)}\int_0^T \sum_{i=1}^{N} \sum_{j\in N_i(t)/i} \frac{v_i(t)v_j(t)}{\|v_i(t)\|\|v_j(t)\|}dt \tag{8.72}$$

where $j \in N_i(t)/i$ means $j \in N_i(t)$ and $j \neq i$. This expression indicates the stability of the flock under different conditions. We did not use the average velocity as the order parameter here, just because velocity correlation (as order parameter) could offer a more strictly stability description when the system is divided into two or more coherently moving subgroups.

(2) **Simulation Results Analysis and Discussion**

We compare VEM with DHM and DHMZ separately. Figure 8.18 shows the comparison results of the VEM and DHM, while Fig. 8.19 shows the comparison results of VEM and DHMZ.

According to Fig. 8.18, we can see that VEM is a little bit better (better meaning that it is more ordered for the same amount of noise) than the DHM for different group size under lower noise, while it seems that DHM is a little better than VEM under larger noise. However, the advantage is so small that can be ignored. At the same time, it seems that there exists no relevant difference among the simulation results for various group sizes.

Figure 8.19 demonstrates that the DHMZ hierarchical model is significantly more stable than an egalitarian model for flocking, and the advantage is increasingly

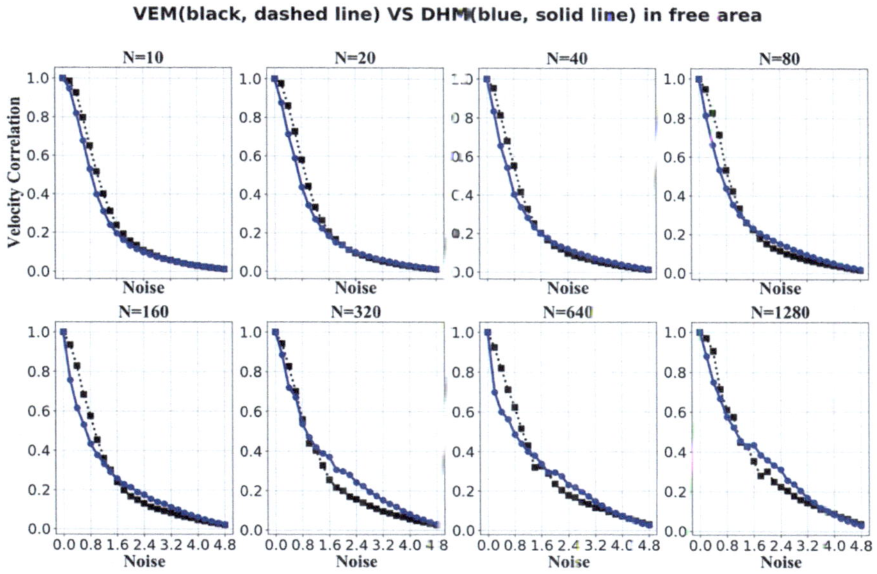

Fig. 8.18 Quantitative comparison of VEM and DHM ($l_{ii} \neq 0, \forall i = 1, ..., N$) for various group sizes. c_{ij} satisfies log-normal distribution (whose mean is 0 and standard deviation is 1) for DHM

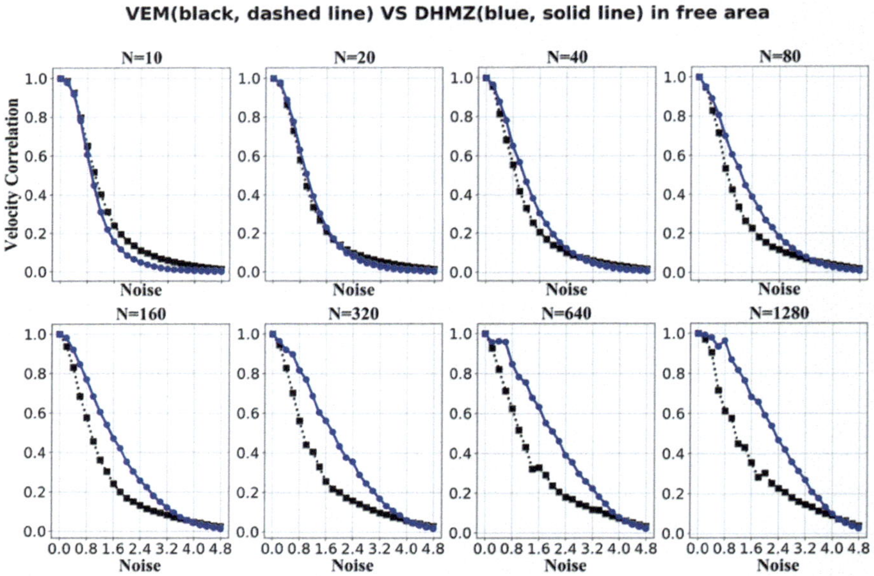

Fig. 8.19 Quantitative comparison of VEM and DHMZ ($l_{ii} = 0$ for $i \in 1, \cdots, N$) for various group sizes. c_{ij} satisfies log-normal distribution(whose mean is 0 and standard deviation is 1) for DHMZ

obvious as the size of the system increases. However, the stability gap between an egalitarian flock and a hierarchical flock stops increasing from 320 to 1280 particles. That is to say, the difference in the performances between the egalitarian system and the hierarchical system is increasing with system size, but only up to a given flock size and is not changing as the group size reaches a certain threshold. This is quite along the intuitive picture which suggests that hierarchy may play a relevant role only up to - at most - a few hundred of flock members.

At first, we were puzzled by the result that DHMZ performs much better than DHM. However there exists a plausible interpretation. The zero on the diagonal means that a unit does not consider its own decision (direction) when it decides about its direction at each time step. The element on the diagonal being not zero means that when a unit makes a decision, it will implement the average decision of both itself and its neighbors. In short, an interaction matrix with zeros along the diagonal means that the "follow the leaders" principle is manifested in an enhanced way. Even if in other interactions the given unit is usually a leader (having a relatively large b_{ij} or c_{ij} values), when this leader decides how to proceed, it becomes a follower in relation with the others.

Let us consider twenty-eight particles, for example, Fig. 8.20 shows some important scenarios during the flocking process. The color bar reveals the weight of the contribution of the given particle. For example, the red particle is the strongest one, while the purple one is the weakest particle. The contribution values of these particles belonging to the DHMZ system satisfy log-normal distribution. That is, $c_{ij}, i, j = 1, ..., N, i > j$ satisfy log-normal distribution, with mean value 0 and standard deviation 1. At the same time, the contribution value of each particle belonging to VEM system is equal, and the sum of the contribution values of VEM system is equal to the sum of the contribution values of the DHMZ system. The initial state of the twenty-eight particles in VEM is the same as that in DHMZ. In Fig. 8.20, the boundary particles are indicated by a bigger size than the inside particles. The first picture in Fig. 8.20 displays the state of all the particles at start. The left part of Fig. 8.20b depicts the structure of the VEM flock being less cohesive than the DHMZ (the right one) and a couple of particles are already separated from the VEM flock. Figure 8.20c and d show two key moments when VEM results in three separated groups while DHMZ results still in a single cluster. This can be taken as an important evidence to demonstrate that leader–follower hierarchical systems are more efficient regarding their stability against perturbations which the individual units are subject to.

8.3.5 *Summary*

In conclusion, we have introduced an interpretation of the possible behaviours of interacting, collectively moving units (let it be birds, fish or people, etc.) which is more general than any of the ones we are aware of. The main feature of our approach is that it unites into a single formalism the results and assumptions of the prior models and observations. Some of our results are plausible: a simple averaging of the direction of motion of neighbours (VEM) is quite efficient when the efficiency is measured as the level of stability against external/internal perturbations. The main points of our paper are concerned with the role of hierarchy among the co-moving units. We have located two non-trivial models: (i) the contribution during the averaging of the directions is weighted, (ii) the given unit is—when evaluating its direction in the net time step—in a fully "follower mode" (DHMZ). At first, our numerical findings (i) is inefficient, (ii) performs the best, looked as ssurprising, but we provided the interpretations for these findings. Our formalismis general enough for using it to introduce a large family of more specific models.

8.4 Optimization of Individual Visual Field

Models have proven useful in revealing the mechanisms underlying these collective behaviors of flocks of birds, schools of fish, swarms of insects, and herds of mammals. Biologists have found that the interaction among the starlings depends on topological rather than metric distance. Most of these topology-based collective models endow single particle with a global visual field, in order to provide each particle with sufficient information about the surroundings. However, in nature, the

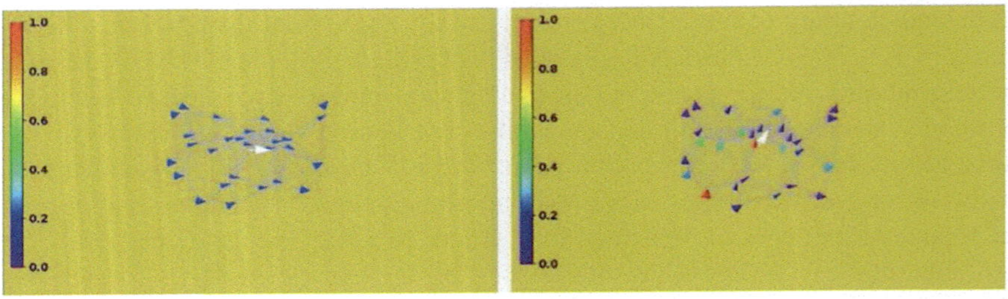

(a) $t = 1s$

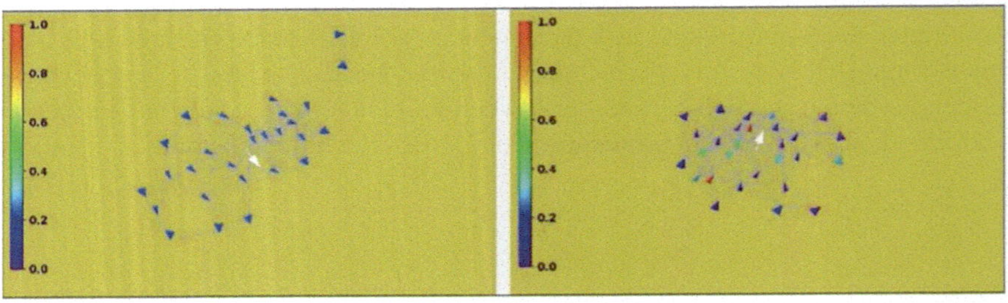

(b) $t = 27s$

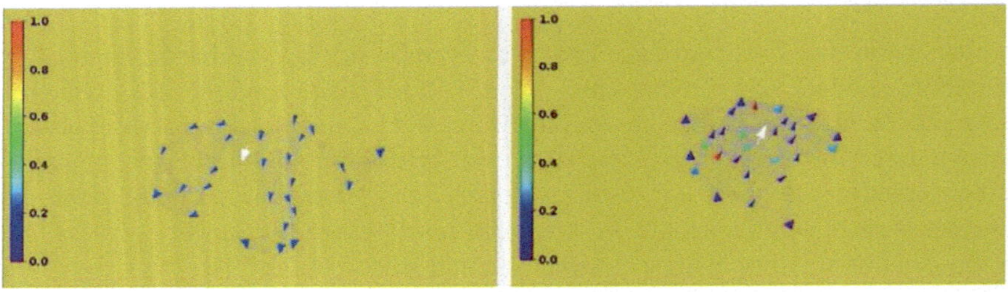

(c) $t = 53s$

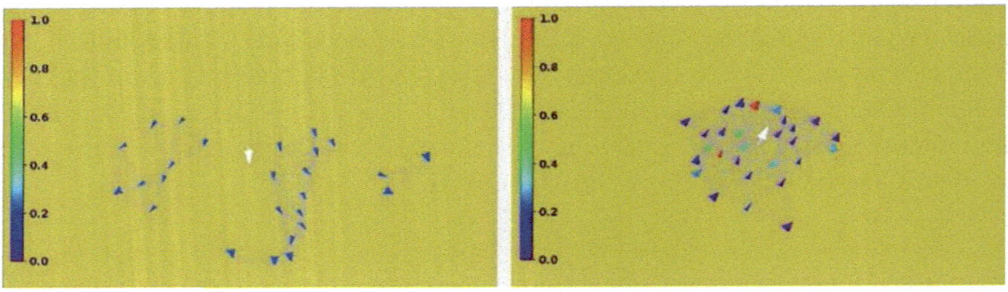

(d) $t = 56s$

Fig. 8.20 VEM and DHMZ for twenty-eight particles flocking. For each scenario, the left one is VEM, while the right one is DHMZ. **a** The scenario at time $t = 1$ s. **b** The scenario at first separate time $t = 27$ s. **c** The scenario at time $t = 53$ s. **d** The scenario at the second separate time $t = 56$ s

individual in a group usually has a restricted visual field, such as the visual field of starlings is 143°. In this paper, we mainly investigate whether a global visual field is necessary to form a consistent and cohesive group in swarms. In order to solve this problem, we propose a three-dimensional restricted visual field metric-free (RVFMF) model based on Pearce and Turner's previous work. Furthermore, we discuss several vital factors governing the convergent consistency of the RVFMF model with the assistance of extensive numerical simulations. According to the simulation results, we conclude that the best view angle of each particle in a swarm increases with increasing population size. Besides, the best view angle gradually becomes stable around 155° when the population size is larger than 1000. We also present data from quantitative analysis to prove that a flock of birds could obtain better coordination pattern under an optimal restricted visual field rather than under a global visual field.

8.4.1 Introduction

Collective intelligence offers effective solutions for groups of organisms during migration, foraging, or predator avoidance, especially for those individuals who cannot survive alone. Scientists are unlocking the secrets to animals' collective behaviour using high-speed video and software to determine how they do it. Models of collective behaviour have proven useful in revealing the mechanisms underlying collective behaivors of flocks of birds, schools of fish, swarms of insects, and herds of mammals.

In 1986, Reynolds introduced the Boid model to depict the emergence of actual collective behaviour of birds in a computer simulation, which has been regarded as the starting point of scientific research on collective models [87]. Since then, a variety of collective models and their inherent properties have been discussed in detail, such as the Vicsek model [88, 89], Cucker-Smale(CS) model [90, 91], hierarchical flocking model [92–94], and metric-free model [95–98].

The Vicsek model, known as the simplest and most classic model, displays a transition from a random state to coordination state [88]. Subsequent numerical studies greatly helped clarify its properties regarding this transition [99–106]. The basic problem of collective behaviour involves how groups of individuals cohere to form a swarm configuration. Most studies either confined the whole swarm to a periodic boundary condition (PBC) [88] or introduced attraction and repulsion terms [107, 108], or some potential field [109]. Nevertheless, Pearce and Turner found that although these approaches can generate swarm cohesion, they also introduced a metric to the system [98]. These methods usually produce feature densities that are virtually constant and independent of the number of individuals in a group [110].

Boid model and Vicsek model have indicated that these collective behaviors of animals can be produced just according to local interaction between individuals. Nevertheless, little is known about the nature of this interaction, so most research results have relied on prior assumptions [111–112]. Ballerini et al. asserted that this interaction does not depend on the metric distance but instead on the topological interaction mode, which is determined by reconstructing the three-dimensional positions of thousands of individual birds in an airborne flock [113]. According to the experimental data, another critical topological feature of bird interactions is that each individual interacts with only up to six or seven neighbours. Besides, Francesco et al. further supposed that these particles in a flock only select neighbours in the first shell of Voronoi tessellation as objects for interaction [96].

Cavagna et al. measured the degree of correlation of the velocity fluctuation of different birds by reconstructing each bird's three-dimensional position and velocity in a large number of starlings, and these experimental data indicated that this behaviour correlation is scale-free [95]. Wolfe et al. proposed a scale-free Vicsek model by employing a new Enskog-type theory, and simulation results proved that the transition from random state to coordination state is continuous in contrast to the case in the traditional metric Vicsek model [97]. In addition, Camperi et al. compared the stability of topological models with that of metric ones, and they concluded that the number of interacting neighbours of each particle is constant and the topological interaction mode is better at avoiding signal interference among particles [114].

Above all, the topological interaction mode seems better than the metric one in a series of aspects. Nevertheless, the swarming density of most models involves topological interactions approaches zero under spatial expansion. In response to this problem, Pearce and Turner proposed a strictly metric-free (SMF) swarming model that introduced a metric-free motional bias to each individual topologically identified as being at the edge of the swarm [98]. In SMF model, the density of these particles can be changed in the unbounded region, and the motorial bias enables the system to achieve an aggregation state in the unbounded three-dimensional space. Furthermore, RVF model is further developed by introducing another outward bias to these birds within the group, besides endowing these birds at the edge of the swarm with an inward movement bias, and they found that the particles in the periphery area are distributed more densely than those in the central area [115].

Couzin et al. believed that each individual in a group uses visual information as the predominant modality to interact with each other when making collective decisions [116, 117]. Most existing collective models suppose that every particle can obtain global vision information [88, 98]. However, in nature, particles in a flock usually have a restricted visual field. For example, the visual field of starlings is 143°, that of pigeons is 158°, while that of owls is 100.5° [118–120]. Although it is difficult to infer how such limited visual fields developed through natural selection over the course of evolution, we can quantitatively analyze the advantages of a restricted visual field compared with a global one.

Based on the above considerations, several studies have been investigated from the point view of understanding the biological mechanism of collective model in nature [121–124], as well as improving the convergence properties of this collective behaviour further [123–130]. For example, Wang et al. focused on minimising the convergence time until reaching a direction consensus in a two-dimensional Vicsek model using exponential neighbour weight and a restricted visual field [127]. In addition, Nguyen et al. studied the effect of visual angle on the phase transition in the two-dimensional Vicsek model. They found that phase transition occurs when the view angle of each particle is more than $\frac{\pi}{2}$ [128]. Durve et al. believed that both the directionality and the radial range of the interaction play essential roles in determining the nature of the phase transition [129]. Moreover, Li et al. optimised the three-dimensional Vicsek model in order to obtain the quickest direction consensus. They found that the optimal view angle is associated with the absolute velocity and the particle density [130].

In this section, a three-dimensional restricted visual field metric-free (RVFMF) model is proposed. Compared with most previous models, the RVFMF model is strictly metric-free and each particle governed by a restricted visual field. The topological neighbor of each particle is defined by these particles that are located in the first shell of Voronoi tessellation as well as within its restricted visual field. Besides, the particles located on the convex hull of the swarm are named as being on the edge, while the others are in the bulk. An edge eigenvalue is introduced to these particles on the convex hull. This eigenvalue introduces a scale-free inward motorial bias for edge particles. "Inward" refers to a vector average pointing from an edge particle to its topological neighbours located on the edge. Edge particles can control the density of the swarm due to the inward motorial bias. When they are treated as a prey, these particles located at the outermost part of the group are the most vulnerable to predation and will spontaneously try to move closer to the group. Meanwhile, these particles in the bulk are relatively safe and maintain its current position and velocity as much as possible. The RVFMF model is used to simulate and quantitatively analyze the evolutionary process of a flock of large-scale particles and explore the influence of visual field from the point view of swarm consistency in this paper. The optimal view angle of each particle is characterized by the group size. Finally, whether a global visual field is necessary for the emergence of a flock is clarified by the assistance of extensive numerical simulations.

8.4.2 Restricted Visual Field Metric-Free (RVFMF) Model

We consider N particles moving at the same speed in a three-dimensional borderless region. The main contribution of the RVFMF model is to consider the view angle of each particle during the evolution process of collective behaviour from an initial disordered state to a final ordered state. At initial time $t = 0$, the positions of N independent particles are randomly distributed in a cube with size L, and the velocity direction of each particle is a random value generated from $[0, 2\pi)$. At the sample time t, the position of a particle i is denoted as r_i^t, the velocity direction is $\underline{v}_i^t$, and $\underline{r}_{i,j}^t$ represents the unit vector pointing from particle i to particle j at time t.

As shown in Fig. 8.21, we define the visual field of each particle as the cone-like scattering column area starting from the particle. Suppose that every particle in the same group possesses the same view angle. Let $\theta \in (0, \pi]$ denote the view angle of particle i. When $\theta_i = \pi$, RVFMF becomes a SMF model. That is, every particle has a global vision.

According to Eqs. (8.73)–(8.76), all particles' positions and velocity directions are updated once every time step. The topological neighbours in the visual field of particle i belong to set B_{iv}, which will be updated at each sample time t. All particles

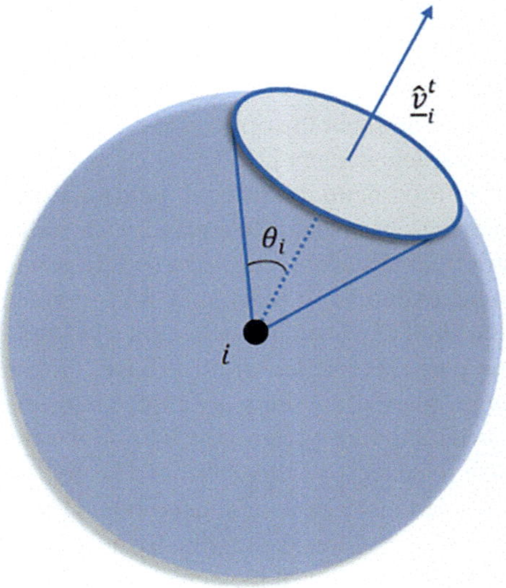

Fig. 8.21 Schematic diagram of the view angle θ of particle *i*

in the convex hull consist of set C. If particle i belongs to set C, then set S_{iv} is equal to all topological neighbors of particle i located on the cluster convex hull, that is, $S_{iv} = B_{iv} \cap C$. This model has two control variables. The first one is edge strength $\emptyset_e$, which is the relative weight of the inward movement deviation of edge particles, and the second one $\emptyset_n$ denotes the intensity of noise

$$\underline{r}_i^{t+1} = \underline{r}_i^t + v_0 \hat{\underline{v}}_i^t \tag{8.73}$$

$$\underline{v}_i^{t+1} = (1 - \emptyset_n)\hat{\underline{u}}_i^t + \emptyset_n \hat{\underline{\eta}}_i^t \tag{8.74}$$

$$\underline{u}_i^t = (1 - f_i)\frac{\hat{\underline{v}}_{jj\in B_{iv}}^t}{\left|\hat{\underline{v}}_{jj\in B_{iv}}^t\right|} \tag{8.75}$$

$$f_i = \begin{cases} \emptyset_e, \underline{r}_i^{t+1} \\ 0, \textit{otherwise} \end{cases} \tag{8.76}$$

$\widehat{\eta_i^t}$ represents a random unit vector at time t for particle i, v_0 represents the speed at which each particle moves, ^ represents the normalized unit vector, and angle brackets < > indicate average subsets. When particle i is located in the bulk, $f_i = 0$, that is, the direction of particle i at the next time step is equal to the average direction of its topological neighbours and has nothing to do with edge strength $\emptyset_e$. When particle i is on the edge of the swarm, $f_i = \emptyset_e$, meaning that the direction of particle i is equal to the linear superposition of the average direction of its topological neighbors multiplied by $(1 - \emptyset_e)$ and the scale-free motorial bias multiplied by $\emptyset_e$.

In order to exhibit the three-dimensional RVFMF model clearly, Fig. 8.22 presents a schematic diagram of the topological constructions in Eqs. (8.74 and 8.75). The purple dots denote the particles $\underline{r}_i^t$ located on the convex hull, while the red dots denote the particles $\underline{r}_i^t$ located in the bulk. To build Voronoi tessellation, Delaunay triangulation must first be performed, as represented by a red line. Point set B_{iv} presents the topological neighbors within the visual field of the target particle, which includes all particles linked to the target particle by red and purple lines. The purple line connects the particles on the convex hull, and in the point set $S_{iv} = B_{iv} \cap C$, pink arrows indicate the scale-free inward motorial bias of these edge particles. Figure 8.23 shows more details of the composition of the metric-free surface term appearing in Eqs. (8.74 and 8.75) for edge particle i. The metric-free surface term of edge particles is equal to the average value of unit vectors ($\hat{\underline{r}}^t_{ijj\in S_{iv}}$) of topological neighbours which are also located on the convex hull, that is, the average of the unit vectors $\hat{\underline{r}}^t_{ia}$ and $\hat{\underline{r}}^t_{ib}$ of adjacent particles a and b. This value has a magnitude of [0, 1] when the angle between particles i and j is small. The longer the arrow is, the larger the value. Therefore, the larger the surface term of the outermost particles in the convex hull is, the stronger the trend of re-entering the swarm. Moreover, to show the convex hull structure and topological neighbor structure of the population more clearly, the structure of the swarm convex hull and the composition of topological neighbors in three-dimensional space are shown in Fig. 8.24. Therein, neighbors located within the field of view of the selected particle are represented by different colors, highlighting the restricted visual field of our model.

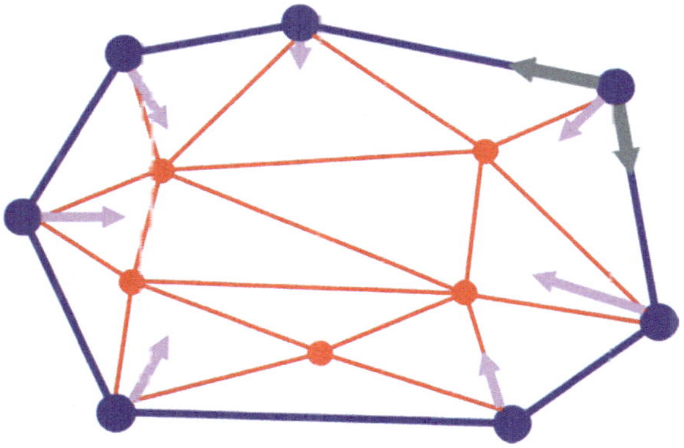

Fig. 8.22 A schematic diagram on the topological constructions of particles

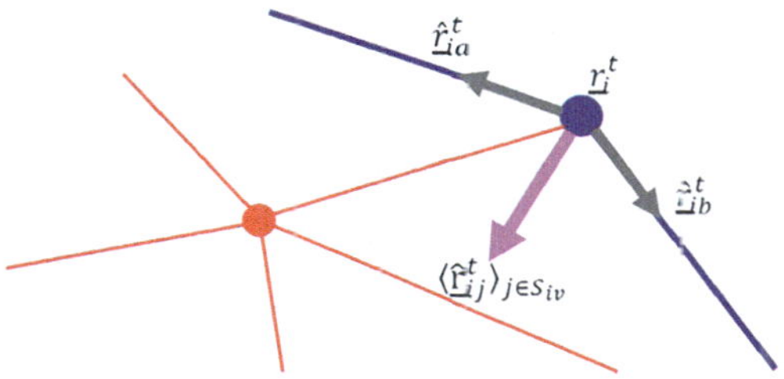

Fig. 8.23 Details on the topological structure of edge particles on convex hull

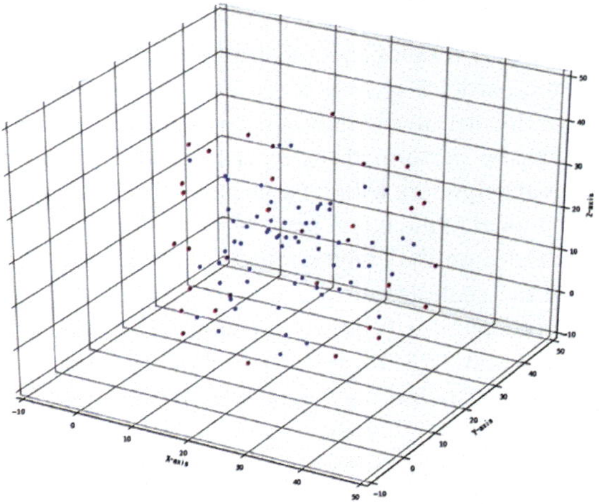

(a)

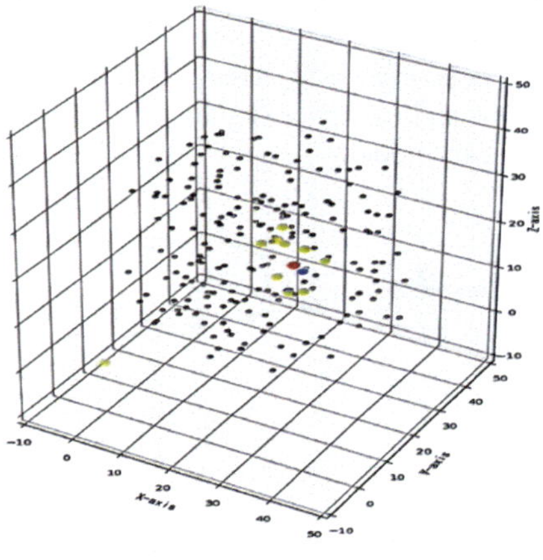

(b)

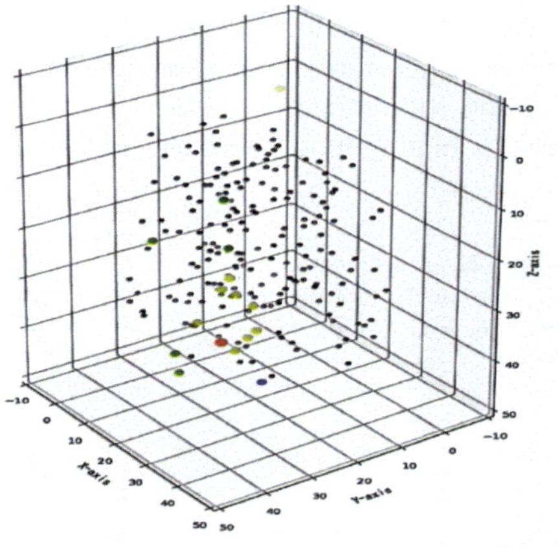

(c)

Fig. 8.24 Convex hull and topological neighbor structure. **a** Convex hull structure: red particles represent individuals on convex hull, blue particles represent individuals in the bulk. **b** Choose red particles randomly, the neighbors outside the field of vision are blue, while the neighbors inside the field of vision are yellow. **c** Topological neighbor structure: red particles on the convex hull are randomly selected, blue particles are their ordinary neighbors, yellow and green particles are their neighbors on convex hull, and yellow ones are the convex hull neighbor inside the visual field

The concept of order degree P is introduced as follows

$$P = \left| \frac{1}{N} \sum_{i=1}^{N} v_i \right| \tag{8.77}$$

which represents the degree of consistency of all individuals in the swarm. The value of P is in the range of [0, 1]. The larger P is, the more consistent these particles are. $P = 1$ denotes the movement direction of all particles is consistent. Similarly, $P = 0$ is obtained, when these particles are randomly distributed at the initial moment.

8.4.3 Simulations Analysis and Results

European starling is a ground-eating bird of the Passeriformes that fly in the daytime. Inspired by its flocking behavior, a three-dimensional collective model called RVFMF is developed. This paper mainly investigates the influence of the visual field of each particle on the stability of the whole system when considering a flock. To exhibit the evolution process of the system from a disordered state to an ordered state, a human–computer interaction interface is built. Figure 8.25 provides an example to present the main interface contents of the simulation platform, from which we can observe the process of change of a cluster from its initial state (as shown in Fig. 8.25a to a stable one (as shown in Fig. 8.25b). The color of each particle denotes its velocity direction. The same color means that the particles are headed in the same direction. Figure 8.25c shows the curve of the global order P of the system over time.

8.4.3.1 Two Control Parameters

One of the main purposes of this paper is to study the influence of the visual field of each particle on stability of the flocking system. To explore this, we need to keep other variables unchanged. Our model has two control parameters: the weight of inward motion bias of particles on the edge $\boldsymbol{\phi}_e$ and the magnitude of noise $\boldsymbol{\phi}_n$.

First, we change the two control parameters separately and keep the other variables unchanged. The goal is to observe the influence of these two parameters on the global order P. Suppose each particle possesses global vision. We observe the changes of the global order of swarm with different group sizes affected by the noise of intensity $\boldsymbol{\phi}_n$ with the simulation results shown in Fig. 8.26. The results show that the swarm can achieve a higher degree of consistency only when there is minor noise interference. The greater the noise is, the lower the consistency of the swarm; a change of the population size does not change this trend. These results are the same as those in a previous report [98], which supports the effectiveness of our model. That is, when each particle has a global vision, the RVFMF model becomes an SMF model. In addition, we can see from Fig. 8.26 that the phase transition occurred when $\boldsymbol{\phi}_n$ around 0.7.

Next, under global vision, the change of the global order of swarms with different group sizes affected by edge strength $\boldsymbol{\phi}_e$ was observed, with Fig. 8.27 presenting the simulation results. The results indicate that, when the population size did not reach 1000, the change of $\boldsymbol{\phi}_e$ had an evident impact on P. However, as the population size continued to increase, the change of $\boldsymbol{\phi}_e$ value had little effect on P. As such, when the

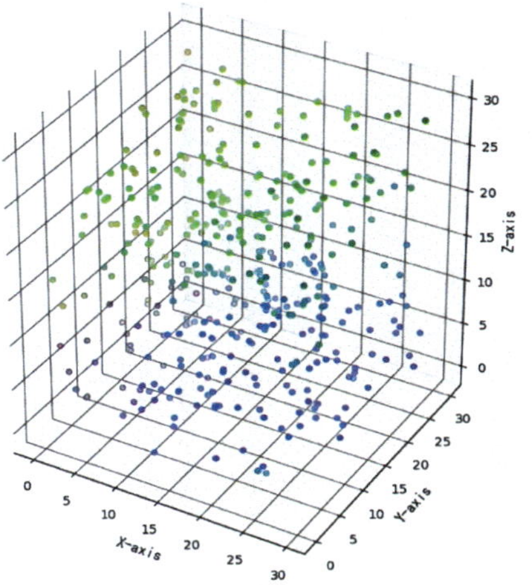

(a)

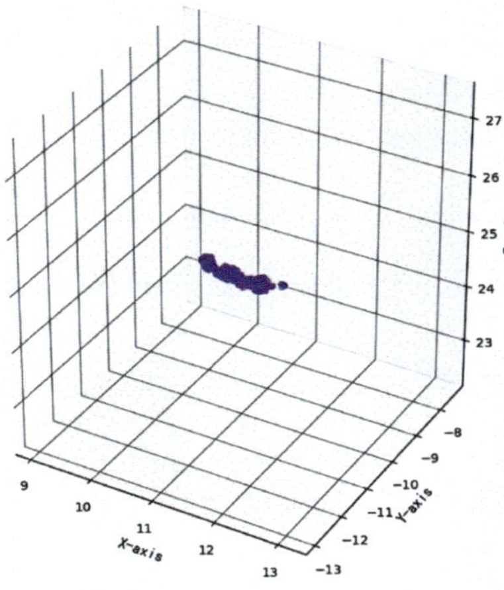

(b)

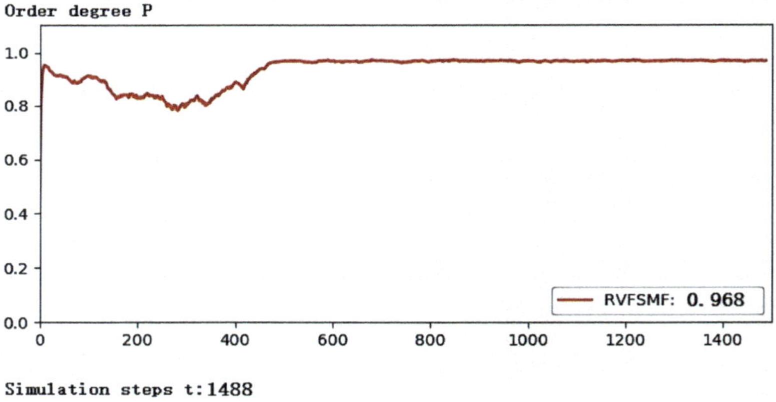

(c)

Fig. 8.25 Swarming from disordered state to ordered state, in which each particle owns a global vision, i.e. $\theta_i = \pi (i = 1, \ldots N)$. Other parameters are population size $N = 400$, noise $\phi_n = 0.2$, edge strength $\phi_e = 0.5$. The color of each particle represents its speed direction. The closer the color is, the higher the global order P of the swarm is. (a) Initial state of random distribution. (b) Stable state of overall speed consistency. (c) Global order P of a flock tends to be stable from 0 to 0.968

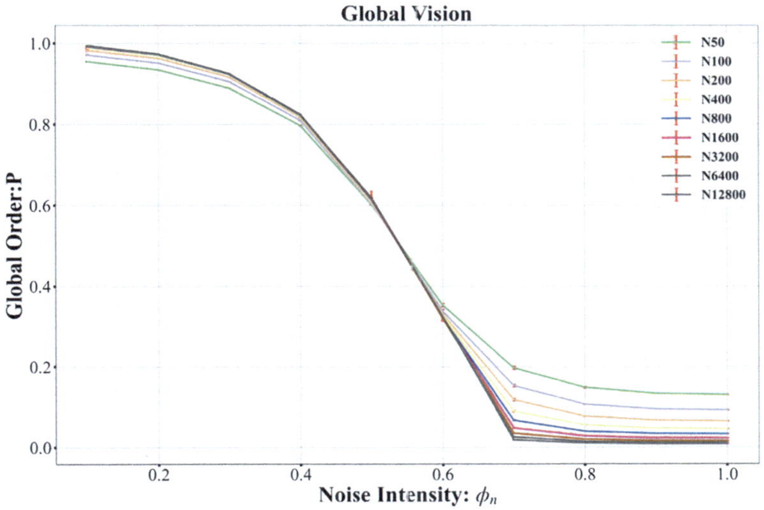

Fig. 8.26 The order parameter P versus the noise strength ϕ_n under various group sizes of particles with global view. Here, $\phi_e = 0.5$

population size is small, the higher the $\boldsymbol{\phi}_e$, the lower the order degree that the swarm can achieve. According to Eq. (8.75), the trends of co-aligning with neighbors and re-entering the swarm are weighted by factors $(1 - \boldsymbol{\phi}_e)$ and $\boldsymbol{\phi}_e$, respectively. The larger the $\boldsymbol{\phi}_e$ value is, the smaller the weight of cooperation, so the smaller the value of global order P. Furthermore, under the same population size and the same noise intensity, the stronger the edge is, the worse the system's stability. When the edge strength is no more than 0.5, the system's stability is almost the same with different population sizes. In contrast, when the edge strength is more than 0.5, under the same edge strength, the larger the group size is, the better the system's stability is. Therefore, we will mainly discuss how to optimize the visual field angle according to the system's stability when the edge strength is more than 0.5.

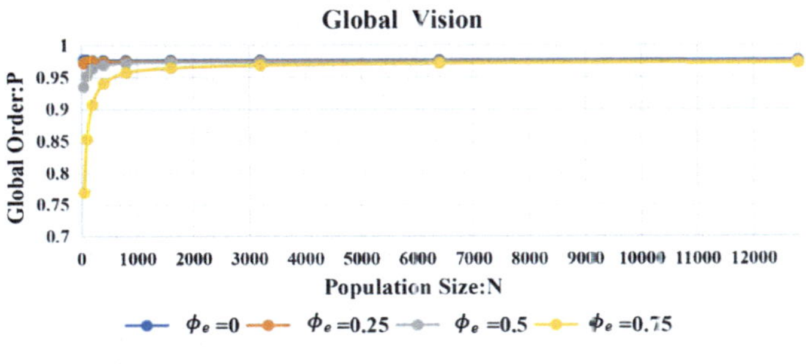

Fig. 8.27 The order parameter P versus the edge strength ϕ_e under various group sizes of particles with global view. Here, $\phi_n = 0.2$

8.4.3.2 Influence of Visual Field Angle

In general, the global angle is considered to provide individuals with the largest field of vision and sufficient information about their surroundings, so the system can achieve the highest degree of cooperation when $\theta = \pi$. However, biological research on European starlings has indicated that the eyeballs of these birds have a specific limited field of vision rather than a global one, during flight, and when starlings are at rest, the visual field is between 130° and 160° in the vertical direction. These features may be determined by complex biological mechanisms. In this paper, we investigate whether the restricted visual field angle benefits the stability of a flock, via quantitative simulation analysis.

Considering that a starling's eyeballs can rotate during flight, which leads to a 20° margin in the vertical direction, we assume that the starling's visual field angle ranges from 110° to 180°. To observe the influence of visual field angle on the order degree of a swarm, we keep other parameters constant and set the visual field angle from 110° to 180° at intervals of 10°.

Based on previous analysis, we would first like to discuss the selection of the edge strength $\boldsymbol{\phi}_e$ and noise intensity $\boldsymbol{\phi}_n$. On the choice of parameter $\boldsymbol{\phi}_n$, we made experiments to make $N = 500$, $\phi_n = 0.2, 0.3$ respectively, we can see how global order changes with different boundary strength($\boldsymbol{\phi}_{\mathrm{n}} = 0.2, 0.3$ can ensure that the noise interference of the swarm is not too small or too large.) By comparing the experiment results, we choose $\phi_n = 0.3$, which can better reflect the change of P value caused by different boundary strength choices compared with the experimental data of $\phi_n = 0.2$. Finally, let the population size be $N = 500$ and the noise intensity $\phi_n = 0.3$. Figure 8.28 shows the curve of global order P under different edge strengths $\boldsymbol{\phi}_e$ and different visual field angles. When $\phi_e = 0.25, 0.35, 0.4$, that is, when $\phi_e \leq 0.5$, the global order P increases with increasing field of view angle θ. This shows that, when the proportion of co-alignment is high, the field of view angle does not appear to have any influence on the starling swarm. In contrast, when $\phi_e > 0.5$, The value of P fluctuates obviously and possesses a maximal value at a certain view angle. By further analyzing the data in Fig. 8.28, the maximum value of P is the most significant when $\phi_e = 0.75$. Therefore, we choose $\phi_e = 0.75$ in the following experiments.

Without loss of generality, we finally choose $\phi_e = 0.75$ and $\phi_n = 0.3$, which may be show results more intuitively if a restricted visual field will give rise to a higher global order than global vision. According to our previous analysis, we keep other parameters constant and set the visual field angle from 110° to 180° at intervals of 10°. In each case, the phase transition process is repeated by 100 times and finally obtains the average value of global order P, which ensure the reliability of the data. In order to further explore the global order in the visual angle intervals between 130° and 150°, we use dichotomy to interpolate between 130° and 150° at intervals of 5°. Finally, we found that under different population sizes, there is a visual angle in the swarm that finally confers the highest global order on the swarm. The results are shown in the Fig. 8.29. We define this angle as the best view angle θ_{opt}. This angle is less than 180°, which is not a global view.

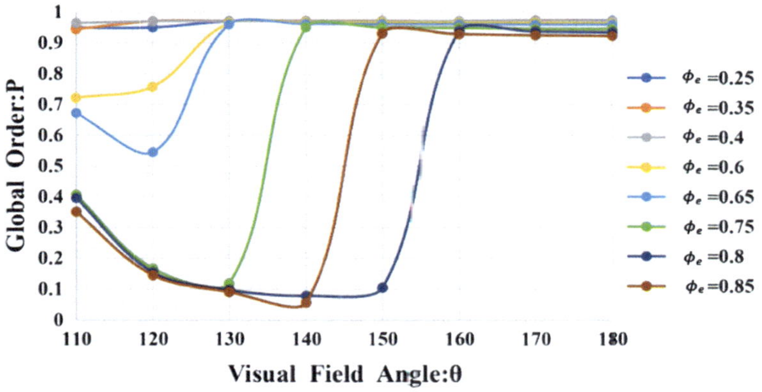

Fig. 8.28 Global order P changes with visual field angle under different values of ϕ_e. Here, $N = 500$ and $\phi_n = 0.3$

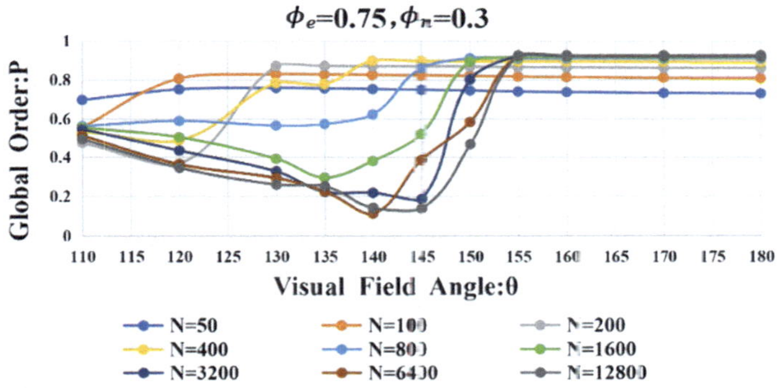

Fig. 8.29 Global order P changes with visual field angle under different group size. When the value of ϕ_e is large, it is difficult for the system to reach a stable state when the number of iteration steps is small. Therefore, after 10,000 pre-balancing steps, the average value of the global order P of 1000 steps is selected as experimental data, and the experiment is repeated 150 times to take the average value

To further explore the relationship between the best view angle θ_{opt} and the population size of the system N. Especially in some population sizes, the global order P is rising in the internal [150°, 160°]. In order to know whether the value of P reaches the maximum within this range, numerous simulation experiments at intervals of 5° were carried out under conditions $\phi_e = 0.75$ and $\phi_n = 0.3$. Figure 8.30 shows the best view angle of the swarm under different population sizes. From the above simulation results, it is concluded that the best view angle of starling swarms increases almost linearly with increasing population size, and θ_{opt} tends to be stable when the population size is more than 1000. It seems that the size of 1000 is a critical inflexion point, which warrants further by study.

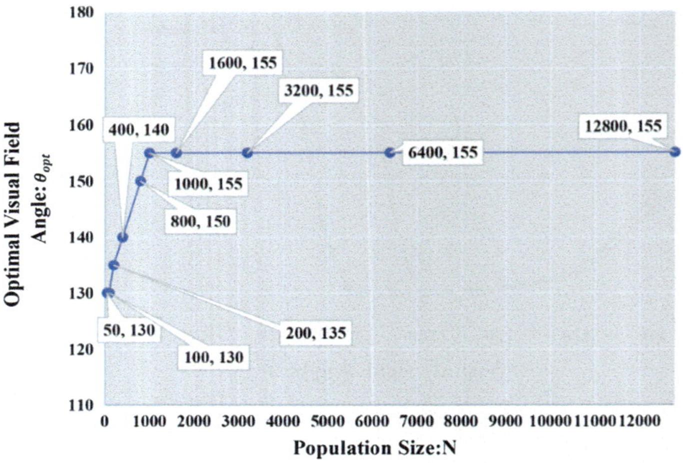

Fig. 8.30 The best view angle changes with the population size. The system conducted 150 experiments under different field of view of each population size, and global order P of each experiment corresponds to an average over 1000 simulation time steps following a 10,000 time step pre-equilibration period

8.4.4 Discussion and Conclusion

In Fig. 8.26, we observe that swarms of different group sizes can achieve the same degree of order parameter under the same degree of noise interference. In nature, it is often observed that a large European starling swarm suddenly divides into two swarms during flight while maintaining the same order state. The results in Fig. 8.26 explain this phenomenon. Under the same degree of noise interference, individuals in the swarm can keep their aggregation state unchanged irrespective of population size. It is obvious that the system is not capable to exhibit coordinate movement when the noise intensity $\boldsymbol{\phi}_n$ is larger than 0.5. With the increase of noise, the swarm undergoes a phase transition from order to disorder.

Figure 8.27 shows that the order degree of the system is associated with the degree of inward bias of individuals who are on the edge of the swarm when the population size is small enough (e.g., ≤ 1000). It is also easy to understand that with the increase of inward bias weight, cooperation is lower, and it is more difficult for swarms to achieve overall consistency of speed. At the same time, we find that, with increasing population size, the influence of ϕ_e becomes increasingly small. That is to say, the larger the size of a population is, the more stable it is, and the less it depends on the edge strength. This explains why European starlings tend to form larger swarms when performing tasks, which are safe and stable.

In the literature, the eyeballs of European starlings are described as having a specific limited field of vision rather than a global visual field during flight. As shown in Fig. 8.28, the experimental results at $\phi_e > 0.5$ are quite consistent with this conclusion. To some extent, the experimental data in Fig. 8.28 also confirm that, in actual starlings, the individuals on the edge of a swarm are more inclined to return to the swarm instead of moving in a manner determined by their neighbors. This tendency effectively guarantees that these individuals do not break away from the swarm.

From the experimental results in Fig. 8.29, we concluded that starlings have an optimal field of view. The experimental results in Fig 8.30 show that the best view angle of the swarm will increase with increasing population size, and that the best view angle tends to converge and stabilise around 155° as the population size reaches around 1000. When the swarm does not reach a threshold (e.g., 1000), individuals need to appropriately increase and adjust their visual field to maintain efficient interactions with increasing population size.

We have got the critical noise value of $\phi_e = 0.5$ from Fig. 8.26. Considering the influence of visual field angle, we would like to know whether this critical value is related to visual angle. Therefore, we added experiments to make the visual field angle equal to 120°, 140°, and 160° respectively, and the experimental results are shown in the Figs. 8.31, 8.32, and 8.33.

According to the above experimental results, the critical noise value has nothing to do with the visual field angle. In further, we reorganized the above data as shown in Fig. 8.34, under each group size, the global order P under view angle 140° and 160° has a higher value than the view angle 120° and 180° under smaller noise intensity ϕ_n, which also implies that a best view angle may exist.

Suppose each particle adopts optimal vision, as shown in Fig 8.30. The changes of the global order of the swarm with different group sizes are affected by noise intensity ϕ_n, as shown in Fig. 8.35. Furthermore, we compared the global order P under global vision with that under optimal vision. Figure 8.36 shows the experimental results. It is clear that when the noise intensity ϕ_n is smaller than 0.5, the system with optimal vision finally approaches higher consistency. Nevertheless, when the noise intensity ϕ_n is more than 0.5, global vision works better. Given that noise intensity ϕ_n is usually

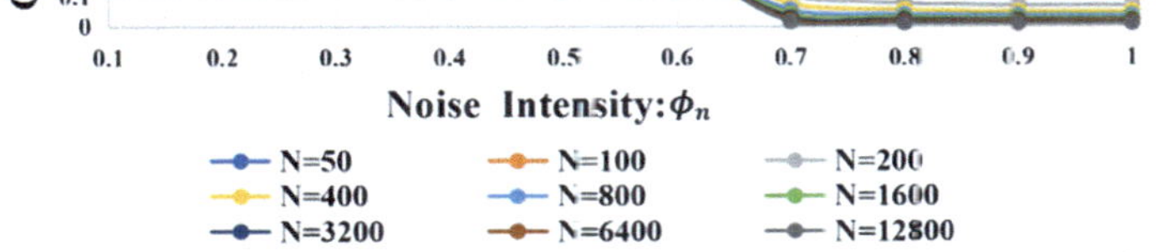

Fig. 8.31 Under the visual field angle of 120°, the swarm is disturbed by different degrees of noise, legend of trend of global order changing with population size. Here, $\phi_e = 0.5$. The same weight of synergy and inward bias is provided to each particle on the boundary

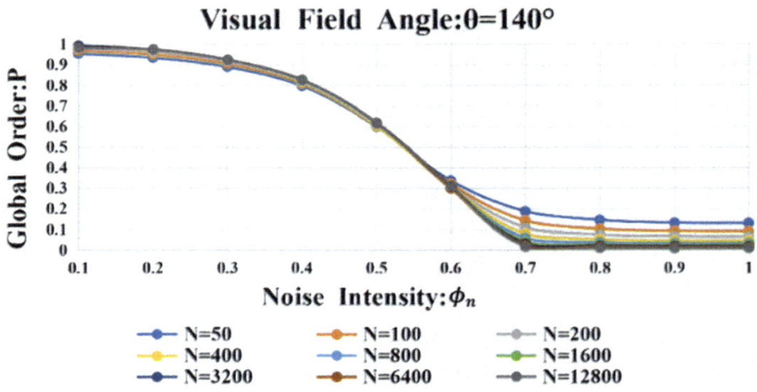

Fig. 8.32 Under the visual field angle of 140°, the swarm is disturbed by different degrees of noise, legend of trend of global order changing with population size. Here, $\phi_e = 0.5$. The same weight of synergy and inward bias is provided to each particle on the boundary

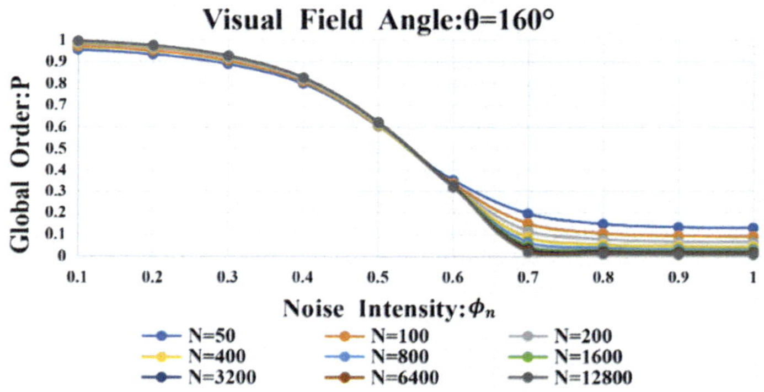

Fig. 8.33 Under the visual field angle of 160°, the swarm is disturbed by different degrees of noise, legend of trend of global order changing with population size. Here, $\phi_e = 0.5$. The same weight of synergy and inward bias is provided to each particle on the boundary

no more than 0.5, individuals tend to evolve an optimal visual field for use during swarm flight, to maintain overall swarm stability, which involves a more complicated biological mechanism. Moreover, optimal vision appears to be a better solution for directional flocking or swarming behaviors.

8.4.5 *Summary*

This paper has investigated the modeling, analysis, and optimization of three-dimensional restricted visual field metric-free(RVFMF) swarm. RVFMF model is used to investigate whether a global visual field is necessary to form a consistent and cohesive group in swarms. Furthermore, statistical methods are adopted to optimize

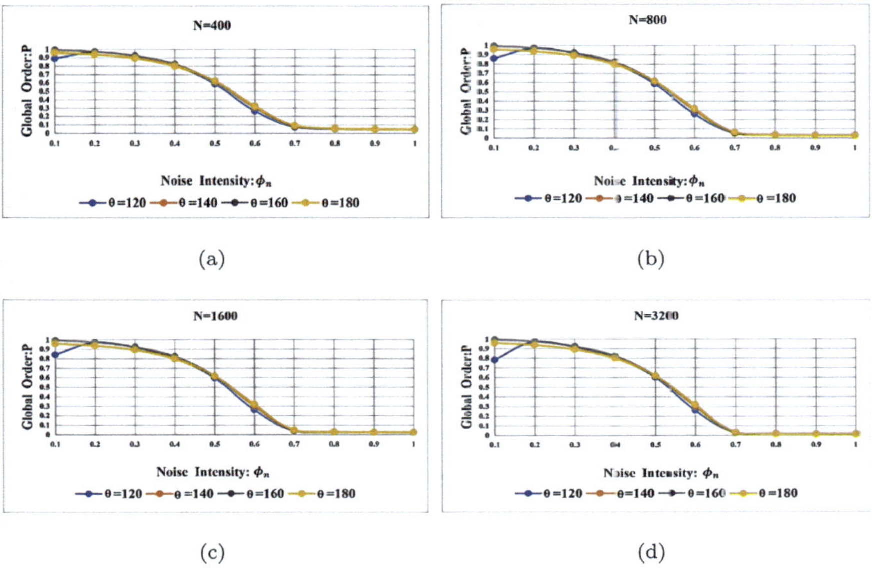

Fig. 8.34 Global order P changed with noise intensity ϕ_n under different angle θ and different population sizes N

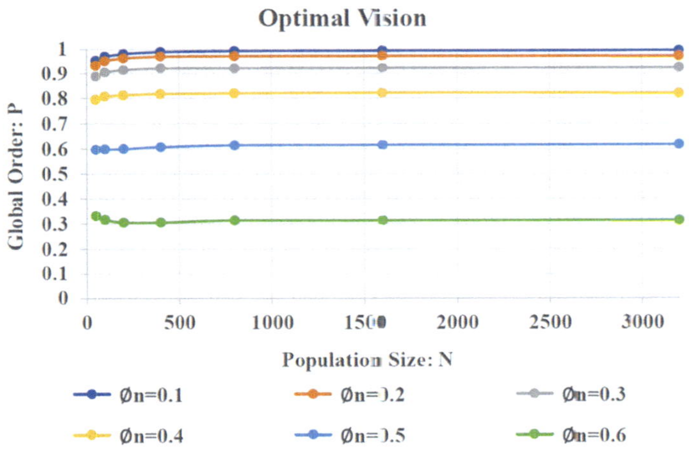

Fig. 8.35 Consider each particle owns the optimal vision angle. The relationship between global order P, noise intensity ϕ_n, and population size N. Here, $\phi_e = 0.75$

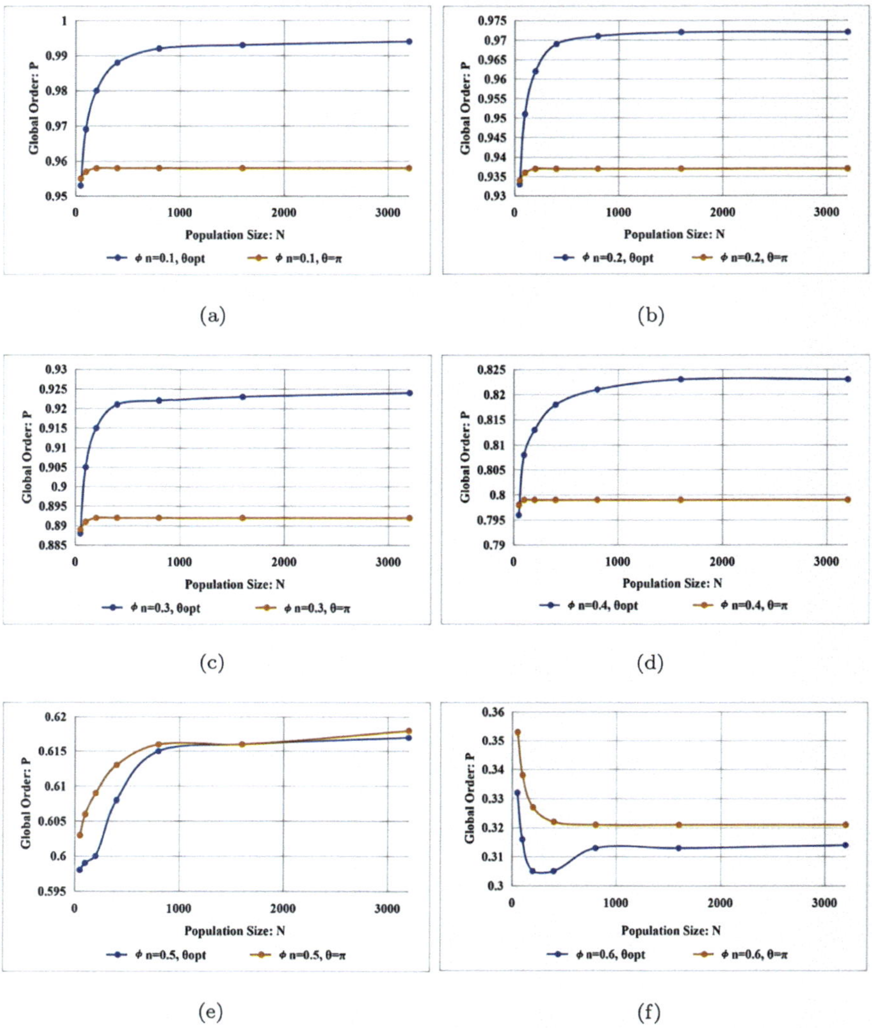

Fig. 8.36 Quantitative comparison on the optimal vision and global vision according to the stability of the swarm system under different noise intensity

the above swarm model relying on several vital factors governing the convergent consistency. According to the simulation results, it is concluded that the best view angle of each particle in a swarm increases with increasing population size. Meanwhile, the best view angle gradually becomes stable around 155° when the population size is larger than 1000. Besides, quantitative analysis is given to prove that a flock of birds could obtain better coordination pattern under an optimal restricted visual field rather than under a global visual field.

References

1. Arenas, A., Diaz-Guilera, A., Kurths, J., Moreno, Y., & Zhou, C. (2008). Synchronization in complex networks. *Physics Reports, 469*, 93–153.
2. Liu, Z. X., & Guo, L. (2009). Synchronization of multi-agent systems without connectivity assumptions. *Automatica, 45*, 2744–2753.
3. Chapman, A., & Mesbahi, M. (2011). UAV flocking with wind gusts: Adaptive topology and model reduction. In *Proceedings of the American Control Conference* (pp. 1045–1050), IEEE.
4. Saska, M., Vakula, J., & Preucil, L. (2014). Swarms of micro aerial vehicles stabilized under a visual relative localization. In *IEEE International Conference on Robotics and Automation* (pp. 3570–3575). IEEE.
5. Reynolds, C. W. (1987). Flocks, herds, and schools: A distributed behavioral model. *Computational Graphics, 21*(4), 25–34.
6. Vicsek, T., Czirook, A., Ben-Jacob, E., Cohen, O., & Shochet, I. (1995). Novel type of phase transitions in a system of self-driven particles. *Physical Review Letters, 75*, 1226–1229.
7. McCune, R., Purta, R., Dobski, M., & Jaworski, A.: Investigations of DDDAS for command and control of UAV swarms with agent-based modelling. In: Proceedings of the Winter Simulations Conference (pp. 1467–1478), IEEE (2013).
8. Buscarino, A., Fortuna, L., Frasca, M., & Rizzo, A. (2006). Dynamical network interactions in distributed control of robots. *Chaos, 16*, 015116.
9. Ben-Asher, Y., Feldman, S., Gurl, P., & Feldman, M. (2008). Distributed decision and control for cooperative uavs using ad hoc communication. *IEEE Transactions on Control Systems Technology, 16*(3), 511–516.
10. Hauert, S., Leven, S., Varga, M., & Ruini, F. (2011). Reynolds flocking in reality with fixed-wing robots: communication range vs maximum turning rate. In *IEEE/RSJ International Conference on Intelligent Robots and Systems.* (pp. 5015–5020), IEEE.
11. Chen, C., Chen, G., & Guo, L. (2015). Consensus of flocks under M-nearest neighbor rules. *Journal of Systems Science and Complexity, 28*, 1–15.
12. Bayezit, I., & Fidan, B. (2013). Distributed cohesive motion control of flight vehicle formations. *IEEE Transactions on Industrial Electronics, 60*(12), 5763–5772.
13. Sun, Y., Li, W., & Zhao, D. (2016). Realization of consensus of multi-agent systems with stochastically mixed interactions. *Chaos, 26*, 073112.
14. Jia, Y. (2016). Swarming coordination of multiple unmanned aerial vehicles in threedimensional space. In *AIAA Modeling and Simulation Technologies Conference*, AIAA.
15. Hung, S. M., & Givigi, S. N. (2017). A q-learning approach to flocking with UAVs in a stochastic environment. *IEEE Trans. Cybern., 47*(1), 186–197.
16. Corner, J. J., & Lamont, G. B. (2004). Parallel simulation of UAV swarm scenarios. In *Proceedings of the Winter Simulation Conference* (pp. 363–371), IEEE.
17. Mendez, L., Givigi, S. N., Schwartz, H. M., Beaulieu, A., Pieris, G., & Fusina, G. (2012). Validation of swarms of robots: Theory and experimental results. In *7th International Conference on System of Systems Engineering* (pp. 332–337), IEEE, Genoa.
18. Ciarletta, L., Guenard, A., Presse, Y., Galtier, V., Song, Y. Q., Ponsart, J. C., Aberkane, S., & Theilliol, D. (2014). Simulation and platform tools to develop safe flock of UAVS: A CPS application-driven research. In *International Conference on Unmanned Aircraft Systems*, (95–102), IEEE.
19. Gil, A. E., Passino, K. M., Ganapathy, S., & Sparks, A. (2008). Cooperative task scheduling for networked uninhabited air vehicles. *IEEE Transactions on Aerospace and Electronic Systems, 4*(2), 561–581.
20. Joelianto, E., & Sagala, A. (2012). Swarm tracking control for flocking of a multi-agent system. In *IEEE Conference on Control, Systems, & Industrial Informatics* (pp. 75–80), IEEE, Bandung.
21. Vasarhelyi, G., Viragh, C., Somorjai, G., Tarcai, N., Szorenyi, T., Nepusz, T., & Vicsek, T. (2014) Outdoor flocking and formation flight with autonomous aerial robots. In *IEEE/RSJ International Conference on Intelligent Robots and Systems*, (pp. 3866–3873), IEEE.

22. Jing, G., Zheng, Y., & Wang, L. (2014). Flocking of multi-agent systems with multiple groups. *International Journal of Control, 87*, 2573–2582.
23. Saska, M. (2015). MAV-swarms: Unmanned aerial vehicles stabilized along a given path using onboard relative localization. In *International Conference on Unmanned Aircraft Systems*, (pp. 894–903), IEEE.
24. Cucker, F., & Smale, S. (2007). Emergent behavior in flocks. *IEEE Transactions on Automatic Control, 52*(5), 852–862.
25. Olfati-Saber, R. (2006). Flocking for multi-agent dynamic systems: Algorithms and theory. *IEEE Transactions on Automatic Control, 51*(3), 401–420.
26. Su, H. S., Wang, X. F., & Chen, G. R. (2009). A connectivity-preserving flocking algorithm for multi-agent systems based only on position measurements. *International Journal of Control, 82*(7), 1334–1343.
27. Vanualailai, J., Sharan, A., & Sharma, B. (2013). A swarm model for planar formations of multiple autonomous unmanned aerial vehicles. In *IEEE International Symposium on Intelligent Control* (pp. 206–211), IEEE, Hyderabad.
28. Beard, R. W., McLain, T. W., Nelson, D. B., Kingston, D., & Johanson, D. (2006). Decentralized cooperative aerial surveillance using fixed-wing miniature UAVs. *Proceedings of the IEEE, 94*(7), 1306–1324.
29. Jadbabaie, A., Lin, J., & Morse, A. S. (2003). Coordination of groups of mobile autonomous agents using nearest neighbor rules. *IEEE Transactions on Automatic Control, 48*(6), 988–1001.
30. Jia, Y., & Zhang, W. (2014). Distributed adaptive flocking of robotic fish system with a leader of bounded unknown input. *International Journal of Control, Automation and Systems, 12*(5), 1049–1058.
31. Wu, S. (2013). *Flight Control System*, 2nd ed. (Beijing University of Aeronautics and Astronautics (BUAA) Press).
32. Wang, L., & Chen, G. (2016). Synchronization of multi-agent systems with metric topological interactions. *Chaos, 26*, 094809.
33. Tanner, H. G., Jadbabaie, A., & Pappas, G. J. (2007). Flocking in fixed and switching networks. *IEEE Transactions on Automatic Control, 52*(5), 863–868.
34. Khalil, H. K. (2002). *Nonlinear systems* (3rd ed.). Prentice Hall.
35. Jia, Y., & Wang, L. (2015). Leader-follower flocking of multiple robotic fish. *IEEE/ASME Transactions on Mechatronics, 20*(3), 1372–1383.
36. Chung, S. J., Paranjape, A. A., Dames, P., Shen, S. J., & Kumar, V. (2018). A survey on aerial swarm robotics. *IEEE Transactions on Robotics, 34*, 837–855.
37. Berrondo, M., & Sandoval, M. (2016). Defining emergence: Learning from flock behavior. *Complexity, 21*, 69–78.
38. Brambilla, M., Ferrante, E., Birattari, M., & Dorigo, M. (2013). Swarm robotics: A review from the swarm engineering perspective. *Swarm Intelligence, 7*, 1–41.
39. Arafat, M. Y., & Moh, S. (2019). Localization and clustering based on swarm intelligence in UAV networks for emergency communications. *IEEE Internet of Things Journal, 6*, 8958–8976.
40. Vasarhelyi, G., Viragh, C., Somorjai, G., Nepusz, T., Eiben, A.E., Vicsek, T. (2018). Optimized flocking of autonomous drones in confined environments. *Science Robotics 3*.
41. Liao, F., Teo, R., Wang, J. L., Dong, X., Lin, F., & Peng, K. (2017). Distributed formation and reconfiguration control of VTOL UAVs. *IEEE Transactions on Control Systems Technology, 25*, 270–277.
42. Moraes, R. S. D., & Freitas, E. P. D. (2017). Distributed control for groups of unmanned aerial vehicles performing surveillance missions and providing relay communication network services. *Journal of Intelligent and Robotic Systems, 3*, 1–12.
43. Tang, Y., Hu, Y., Cui, J., Liao, F., Lao, M., Lin, F., & Teo, R. S. H. (2019). Vision-aided multi-UAV autonomous flocking in GPS-denied environment. *IEEE Transactions on Industrial Electronics, 66*, 616–626.

44. Mukherjee, A., Misraa, S., Sukrutha, A., & Raghuwanshic, N. S. (2020). Distributed aerial processing for IoT-based edge UAV swarms in smart farming. *Computer Networks, 167*, 107038.
45. Olfati-Saber, R., & Murray, R. M. (2004). Consensus problems in networks of agents with switching topology and time-delays. *IEEE Transactions on Automatic Control, 49*, 1520–1533.
46. Olfati-Saber, R. (2006). Flocking for multi-agent dynamic systems: Algorithms and theory. *IEEE Transactions on Automatic Control, 51*, 401–420.
47. Tanner, H. G., Jadbabaie, A., & Pappas, G. J. (2007). Flocking in fixed and switching networks. *IEEE Transactions on Automatic Control, 52*, 863–868.
48. Su, H., Wang, X. F., & Chen, G. R. (2010). Rendezvous of multiple mobile agents with preserved network connectivity. *Systems and Control Letters, 59*, 313–322.
49. Jia, Y., & Wang, L. (2015). Leader-follower flocking of multiple robotic fish. *IEEE/ASME Transactions on Mechatronics, 20*, 1372–1383.
50. Yang, Z. Q., Zhang, Q., & Chen, Z. Q. (2018). Distributed Constraint Optimization with Flocking Behavior. *Complexity, 2018*, 1579865.
51. Moshtagh, N., & Jadbabaie, A. (2007). Distributed geodesic control laws for flocking of nonholonomic agents. *IEEE Transactions on Automatic Control, 52*, 681–686.
52. Jing, G., Zheng, Y., & Wang, L. (2014). Group flocking of multiple mobile agents. In *Proceedings of the 33rd Chinese Control Conference* (pp. 1155–1161), Nanjing, China.
53. Jadbabaie, A., Lin, J., & Morse, A. S. (2002). Coordination of groups of mobile autonomous agents using nearest neighbor rules. *IEEE Transactions on Automatic Control, 48*, 988–1001.
54. Gu, D., & Wang, Z. (2009). Leader-follower flocking: Algorithms and experiments. *IEEE Transactions on Control Systems Technology, 17*, 1211–1219.
55. Jia, Y., & Zhang, W. (2014). Distributed adaptive flocking of robotic fish system with a leader of bounded unknown input. *International Journal of Control Automation and Systems, 12*, 1049–1058.
56. Jia, Y. (2016) Swarming coordination of multiple unmanned aerial vehicles in three-dimensional space. In *Proceedings of the AIAA AVIATION Forum: Modeling and Simulation Technologies Conference*, Washington, DC, USA, 13–17 June 2016.
57. Ballerini, M., Cabibbo, N., Candelier, R., Cavagna, A., Cisbani, E., Giardina, I., Lecomte, V., Orlandi, A., Parisi, G., Procaccini, A., et al. (2008). Interaction ruling animal collective behavior depends on topological rather than metric distance: Evidence from a field study. *Proceedings of the National academy of Sciences of the United States of America, 105*, 1232–1237.
58. Rybski, P., Larson, A., Lindahl, M., & Gini, M. (1998). Performance evaluation of multiple robots in a search and retrieval task. In *Proceedings of the Artificial Intelligence and Manufacturing Workshop* (pp. 153–160), AAAI Press, Menlo Park, CA, USA
59. Rosenfeld, A., Kaminka, G. A., & Kraus, S. (2006). A study of scalability properties in robotic teams. In *Coordination of Large-scale Multiagent Systems Part I*; Springer: Berlin/Heidelberg, Germany.
60. Timothy, S., & Dario, F. (2010). Energy-time efficiency in aerial swarm deployment. *Springer Tracts Adv Robot 83*
61. Wang, Z., Liu, L., & Long, T. (2017). Minimum-time trajectory planning for multi-unmanned-aerial-vehicle cooperation using sequential convex programming. *Journal of Guidance, Control, and Dynamics, 40*, 1–7.
62. Dai, R., & Cochran, J. E. (2009). Three-dimensional trajectory optimization in constrained airspace. *Journal of Aircraft, 46*, 627–634.
63. Jia, Y., & Wang, L. (2014). Experimental implementation of distributed flocking algorithm for multiple robotic fish. *Control Engineering Practice, 30*, 1–11.
64. Reynolds, C. (1987). Flocks, herds, and schools: A distributed behavioral model. *Computational Graphics, 21*, 25–34.
65. Gazi, V., & Passino, K. M. (2003). Stability analysis of swarms. *IEEE Transactions on Automatic Control, 48*, 692–696.

66. Olfati-Saber, R., Fax, J. A., & Murray, R. M. (2007). Consensus and Cooperation in Networked Multi-Agent Systems. *Proceedings of the IEEE, 95*, 215–233.
67. Sumpter, D. J. T. (2010). *Collective animal behavior*. Princeton University Press.
68. Nagy, M., Akos, Z., Biro, D., & Vicsek, T. (2010). Hierarchical group dynamics in pigeon flocks Nature, 464890–3.
69. Vicsek, T., & Zafeiris, A. (2012). Collective motion. *Physics Reports, 517*, 71–140.
70. Petit, O., & Bon, R. (2010). Decision-making processes: The case of collective movements Behav. *Process., 84*, 635–647.
71. Ozogany, K., & Vicsek, T. (2015). Modeling the emergence of modular leadership hierarchy during the collective motion of herds made of haremsJ. *Stat. Phys., 158*, 628–646.
72. King, A. J., Douglas, C. M. S., Huchard, E., Isaac, N. J. B., & Cowlishaw, G. (2008). Dominance and affiliation mediate despotism in a social primate Curr. *O Biologico, 18*, 1833–1838.
73. Seeley, T. D. (2010). *Honeybee democracy*. Princeton University Press.
74. Zafeiris, A., & Vicsek, T. (2013). Group performance is maximized by hierarchical competence distribution Nat. *Commun., 4*, 2484.
75. Vasarhelyi, G., Viragh, C., Somorjai, G., Nepusz, T., Eiben, A. E., & Vicsek, T. (2018). Optimized flocking of autonomous drones in confined environments *Science Robotics 3*, eaat3536
76. Floreano, D., & Wood, R. J. (2015). Science, technology and the future of small autonomous drones. *Nature 521*, 460–C466.
77. Brambilla, M., Ferrante, E., Birattari, M., & Dorigo, M. (2013). *Swarm robotics: A review from the swarm engineering perspective Swarm Intell, 7*, 1–41.
78. Zafeiris, A., & Vicsek, T. (2018). *Why we live in hierarchies? A Quantitative treatise*. Springer.
79. Couzin, I. D., Krause, J., Franks, N. R., & Levin, S. A. (2005). Effective leadership and decision making in animal groups on the move. *Nature, 433*, 513–516.
80. Sueur, C., & Petit, O. (2008). Organization of group members at departure is driven by social structure in macaca. *International Journal of Primatology., 29*, 1085–1098.
81. Sarova, R., Spinka, M., Panama, J. L. A., & Simecek, P. (2010). Graded leadership by dominant animals in a herd of female beef cattle on pasture. *Animal Behavior, 79*, 1037–1045.
82. Nagy, M., Vasarhelyi, G., Pettit, B., Roberts-Mariani, I., Vicsek, T., & Biro, D. (2013). Context-dependent hierarchies in pigeons. *Proceedings of the National Academy of Sciences, 110*, 13049–13054.
83. Shen, J. (2007/08). Cucker-Smale flocking under hierarchical leadership. *SIAM Journal on Applied Mathematics, 68*, 694–719.
84. Pearce, D. J. G., & Turner, M. S. (2014). Density regulation in strictly metric-free swarms New. *Journal of Physics, 16*, 07.
85. Vicsek, T., Czirok, A., Ben-Jacob, E., Cohen, I., & Shochet, O. (1995). Novel type of phase transition in a system of self-driven particles. *Physical Review Letters, 75*, 1226.
86. Reynolds, C. W. (1987). Flocks, herds, and schools: A distributed behavioral model. *ACM SIGGRAPH Computer Graphics, 21*(4), 25–34.
87. Vicsek, T., Czirok, A., Ben-Jacob, E., Cohen, I., & Shochet, O. (1995). Novel type of phase transition in a system of self-driven particles. *Physical Review Letters, 75*(6), 1226.
88. Puzzo, M. L. R., De Virgiliis, A., & Grigera, T. S. (2019). Self-propelled vicsek particles at low speed and low density. *Physical Review E, 99*(5), 052602.
89. Barbaro, A. B. T., Canizo, J. A., Carrillo, J. A., & Degond, P. (2016). Phase transitions in a kinetic flocking model of cucker-smale type. *Multi-scale Modeling and Simulation, 14*(3), 1063–1088.
90. Couzin, I. D., Krause, J., Franks, N. R., & L. S. A. (2005). Effective leadership and decision-making in animal groups on the move, *Nature 433*, 513–516.
91. Nagy, M., Akos, Z., Biro, D., & Vicsek, T. (2010). Hierarchical group dynamics in pigeon flocks. *Nature, 464*(7290), 890.
92. Jia, Y. N., & V. T. (2019). Modelling hierarchical flocking. *New Journal of Physics, 21* (2019) 093048.

93. Cavagna, A., Cimarellib, A., Giardina, I., Parisi, G., Santagatib, R., Stefanini, F., & Viale, M. (2010). Scale-free correlations in starling flocks. *Proceedings of the National Academy of Sciences, 107*(26), 11865–11870.
94. Ginelli, F., & Chate, H. (2010). Relevance of metric-free interactions in flocking phenomena. *Physical Review Letters, 105*(16), 168103.
95. Chou, Y. L., Wolfe, R., & Ihle, T. (2012). Kinetic theory for systems of self-propelled particles with metric-free interactions. *Physical Review E Statal Nonlinear and Soft Matter Physics, 86*, 021120.
96. Pearce, D. J. G., & Turner, M. S. (2014). Density regulation in strictly metric-free swarms. *New Journal of Physics, 16*(8), 082002.
97. Aldana, M., & Huepe, C. (2003). Phase transitions in self-driven many-particle systems and related non-equilibrium models: A network approach. *Journal of Statistical Physics, 112*(1–2), 135–153.
98. Savkin, A. V. (2004). Coordinated collective motion of groups of autonomous mobile robots: Analysis of vicsek's model. *IEEE Transactions on Automatic Control, 49*(6), 981–982.
99. Czirk, A., & Vicsek, T. (2012). Collective behavior of interacting self-propelled particles. *Physica A: Statistical Mechanics and Its Applications, 281*(1–4), 17–29.
100. Barbaro, A. B. T., & Degond, P. (2014). Phase transition and diffusion among socially interacting self-propelled agents. *Discrete and Continuous Dynamical Systems-Series B, 19*(5), 1249–1278.
101. Solon, A. P., Chate, H., & Tailleur, J. (2015). From phase to microphase separation in flocking models: The essential role of nonequilibrium fluctuations. *Physical Review Letters, 114*(6), 068101.
102. Pattanayak, S., & Mishra, S. (2018). Collection of polar self-propelled particles with a modified alignment interaction. *Journal of Physics Communications, 2*(4), 045007.
103. Escaff, D., & Delpiano, R. (2020). Flocking transition within the framework of kuramoto paradigm for synchronization: Clustering and the role of the range of interaction. *Chaos, 30*(8), 083137.
104. Degond, P., Diez, A., Frouvelle, A., & Merino-Aceituno, S. (2020). Phase transitions and macroscopic limits in a bgk model of body-attitude coordination. *Journal of Nonlinear Science, 30*(6), 2671–2736.
105. Chen, Y. H. (2010). Adaptive robust control of artificial swarm systems. *Applied Mathematics and Computation, 217*(3), 980–987.
106. Moussa, N., Tarras, I., Mazroui, M., & Boughaleb, Y. (2011). Effects of agent's repulsion in 2d flocking models. *International Journal of Modern Physics C, 22*(7), 661–668.
107. Bhattacharya, K., & Vicsek, T. (2010). Collective decision making in cohesive flocks. *New Journal of Physics, 12*(9), 093019.
108. Ballerini, M., Cabibbo, N., Candelier, R., Cavagna, A., Cisbani, E., Giardina, I., Orlandi, P. G., & Procaccini, V. (2008). Empirical investigation of starling flocks: A benchmark study in collective animal behaviour. *Animal Behavior London- Baillere Tindall Then Academic Press, 76*, 201–215.
109. Olfati-Saber, R., & Murray, R. M. (2004). Consensus problems in networks of agents with switching topology and time-delays. *IEEE Transactions on Automatic Control, 49*(9), 1520–1533.
110. Chate, H., Ginelli, F., Gregoire, G., Peruani, F., & Raynaud, F. (2008). Modeling collective motion: Variations on the vicsek model. *European Physical Journal B, 64*(3–4), 451–456.
111. Ballerini, M., Cabibbo, N., Candelier, R., Cavagna, A., Cisbani, E., Giardina, I., Lecomte, V., Orlandi, A., Parisi, G., Procaccini, A., Viale, M., & Z. V. (2008). Interaction ruling animal collective behavior depends on topological rather than metric distance: Evidence from a field study. *Proceedings of the National Academy of Sciences 105*(4), 1232–1237 (2008)
112. Camperi, M., Cavagna, A., & Giardina, I. (2012). Spatially balanced topological interaction grants optimal cohesion in flocking models. *Interface Focus, 2*(6), 715–725.
113. Lewis, J. M., & Turner, M. S. (2017). Density distributions and depth in flocks. *Journal of Physics D Applied Physics, 50*(49), 494003.

114. Strandburg-Peshkin, A., Twomey, C. R., Bode, N. W. F., Kao, A. B., Katz, Y., Ioannou, C. C., Rosenthal, S. B., Torney, C. J., Wu, H. S., Levin, S. A., & Couzin, I. D. (2013). Visual sensory networks and effective information transfer in animal groups. *Current Biology, 23*(17), 709–711.
115. Davidson, J. D., Sosna, M. M. G., Twomey, C. R., Sridhar, V., Leblanc, S., & Couzin, I. D. (2021). Collective detection based on visual information in animal groups. *Journal of the Royal Society Interface* https://doi.org/10.1101/2021.02.18. 431380.
116. Couzin, I., Krause, J., & James, R. (2002). Collective memory and spatial sorting in animal groups. *Journal of Theoretical Biology, 218*(1), 1–11.
117. Martin, G. R. (1994). Visual fields in woodcocks scolopax rusticola (scolopacidae; charadriiformes). *Journal of Comparative Physiology A, 174*(6), 787–793.
118. Andreas, H., & Christian, W. (1992). The simulation of the movement of fish schools. *Journal of Theoretical Biology, 156*(3), 365–385.
119. Miguel, M. C., Parley, J. T., & Pastor-Satorras, R. (2018). Effects of heterogeneous social interactions on flocking dynamics. *Physical Review Letters, 120*(6), 068303.
120. Netzer, G., Yarom, Y., & Ariel, G. (2019). Heterogeneous populations in a network model of collective motion. *Physica A: Statistical Mechanics and Its Applications, 530*, 121550.
121. Degond, P., & Merino-Aceituno, S. (2020). Nematic alignment of self-propelled particles: From particle to macroscopic dynamics. *Mathematical Models and Methods in Applied Sciences, 30*(10), 1935–1986.
122. Afsharizand, B., Chaghoei, P. H., Kordbacheh, A. A., Trufanov, A., & Jafari, G. (2020). Market of stocks during crisis looks like a flock of birds. *Entropy, 22*(9), 1038.
123. George, M., & Ghose, D. (2012). Reducing convergence times of self-propelled swarms via modified nearest neighbor rules. *Physica A: Statistical Mechanics and Its Applications, 391*(16), 4121–4127.
124. Chen, G., Liu, Z. X., & Guo, L. (2012). The smallest possible interaction radius for flock synchronization. *SIAM Journal on Control and Optimization, 50*(4), 1950–1970.
125. Wang, X. G., Zhu, C. P., Yin, C. Y., Hu, D. S., & Yan, Z. J. (2013). A modified vicsek model for self-propelled agents with exponential neighbor weight and restricted visual field. *Physica A: Statistical Mechanics and Its Applications, 392*(10), 2398–2405.
126. Nguyen, P. T., Lee, S. H., & Ngo, V. T. (2015). Effect of vision angle on the phase transition in flocking behavior of animal groups. *Physical Review E, 92*(3), 032716.
127. Durve, M., & Sayeed, A. (2016). First-order phase transition in a model of selfpropelled particles with variable angular range of interaction. *Physical Review E, 93*(5), 052115.
128. Li, Y. J., Wang, S., Han, Z. L., Tian, B. M., Xi, Z. D., & Wang, B. H. (2011). Optimal view angle in the three-dimensional self-propelled particle model. *Europhysics Letters, 93*(6), 68003.
129. Jia, Y. N., Li, Q., & Zhang, Z. L. (2020). Accelerating emergence of aerial swarm. *Applied Science, 10*, 7986.
130. Martin, G. R. (1986). The eye of a passeriform bird, the European starling (sturnus vulgaris): Eye movement amplitude, visual fields and schematic optics. *Journal of Comparative Physiology A, 159*(4), 545–557.

The manufacturer's authorised representative in the EU is Springer
Nature Customer Service Centre GmbH, Europaplatz 3, 69115 Heidelberg,
Germany. If you have any concerns regarding our products, please
contact ProductSafety@springernature.com

Printed and bound by CPI Group (UK) Ltd, Croydon, CR0 4YY
01/10/2026
02225945-0019